D0049251

NEW WAYS
TO USE
TEST METERS

A MODERN GUIDE
TO ELECTRONIC
SERVICING

by

Robert G. Middleton

Prentice-Hall, Inc.
Business and Professional Division
Englewood Cliffs, New Jersey

© 1983 by

Prentice-Hall, Inc.
Englewood Cliffs, NJ

Prentice-Hall International, Inc., *London*
Prentice-Hall of Australia, Pty. Ltd., *Sydney*
Prentice-Hall Canada, Inc., *Toronto*
Prentice-Hall of India Private Ltd., *New Delhi*
Prentice-Hall of Japan, Inc., *Tokyo*
Prentice-Hall of Southeast Asia Pte. Ltd., *Singapore*
Whitehall Books, Ltd., *Wellington, New Zealand*
Editora Prentice-Hall do Brasil, Ltda., *Rio de Janeiro*

Editor: George E. Parker

Library of Congress Cataloging in Publication Data

Middleton, Robert Gordon
  New ways to use test meters.

  Includes index.
  1. Electronic apparatus and appliances—
Maintenance and repair.   2. Electronic
measurements.   I. Title.
TK7870.2.M53   1983   621.3815'48   83-3348
ISBN 0-13-616169-3

Printed in the United States of America

A WORD FROM THE AUTHOR

This *practical* answer book for the professional electronic troubleshooter shows and tells the reader how to cope with the new challenges of both digital and analog circuitry. Accordingly, the two chief sections in this volume are:

1. Analog Troubleshooting Techniques
2. Digital Troubleshooting Techniques

The illustrations and descriptions explain how to solve puzzling electronic troubleshooting problems by asking the right questions, measuring the right parameters, and drawing the right conclusions.

Many never-before-published troubleshooting techniques are explained in this book, accompanied by case histories showing how the new techniques are applied, step-by-step. In both the analog and digital sections, these unique troubleshooting techniques include:

- Finger test for identification of transistor terminals
- Constant current gain measurements
- Distortion checking with ac voltmeter
- Self-oscillation mapout technique
- Regeneration mapout technique
- Dynamic internal resistance tests
- Dynamic internal impedance tests
- Troubleshooting with reference clamps
- Forward-biased signal-tracing probe
- Transresistance tests
- Peak-reading probe for very narrow pulses
- Troubleshooting with a peak-hold unit
- Bridge-type probe for distortion tests
- Constant-current transconductance tests
- Amplitude modulation measurements
- Amplifier lock-up and lock-down analysis
- Phase angle measurement with ac voltmeter
- Measurement of dB negative feedback
- Complex waveform checkout with ac voltmeter

- Troubleshooting with a dip-hold unit
- FM/CW identifier probe
- Two-tone digital logic probe
- Coincidence and anti-coincidence probes

You will find this troubleshooting guide to be a most useful reference book. The text explains why particular trouble symptoms are analyzed by making selected tests and measurements—and where new kinds of tests and measurements provide more effective troubleshooting. In many malfunction situations, there are several ways of tackling a puzzling trouble symptom, so what-to-do and how-to-do-it information is supplemented with "why it is done" and "new and better ways to do it" supporting discussion.

Many illustrative examples, specific guidelines, real-life case histories, and tricks of the trade are provided to make the material more understandable and easy to read. For example, the text provides charts for appropriate application fields for the VOM, TVM, and DVM. As an illustration, only a VOM (with its different range voltages) is suitable for checking out the nonlinear resistance characteristic of a diode or transistor. Accordingly, only a VOM is appropriate for precise matching of diodes or transistors. Only a TVM is suitable for "beating out" fundamentals and harmonics in high-impedance circuitry. Only a DVM (in the service-type instrument category) can indicate small amounts of distortion in high-fidelity equipment.

Illustrative examples of specific guidelines are provided. For instance, typical measurements of nonlinear resistance characteristics are cited to clearly show how a VOM provides more precise matching of diodes or transistors than does a TVM or DVM. Case histories are included to underscore generalized test and measurement techniques. For example, a typical "bad-level" trouble symptom, wherein the VOM pinpoints an open bond (IC internal connection) as the cause of the malfunction, is followed from start to finish. Case histories zero in on effective ways to adapt and to easily use the functions of a particular instrument.

Calculator-aided troubleshooting is one of the new troubleshooting methods. In the past, nearly all reactance and impedance calculations were avoided in troubleshooting procedures—few technicians were disposed to make any but the simplest pencil-and-paper calculations. Today the pocket calculator has changed all that—we can now punch out a dynamic internal resistance value or an average/p-p value on the calculator and get answers with a minimum of time and effort. Impedance and reactance measurements, by the same token, are no longer the sole domain of the elite engineer—impedance and reactance values can now be easily solved by the electronic troubleshooter.

In the good old days there were various tragic troubleshooting foul-ups—often days of wasted time and effort—because of instrument limitations and lack of knowledge about circuit functions and malfunctions. Some of the fiascos resulted from failure to supplement resistance measurements with reactance measurements or impedance measurements. Others resulted from oversights concerning the disturbing action of unsuspected feedback waveforms. Still others resulted from failure to take into account the internal resistance of high-voltage/low-current circuits. The modern troubleshooter has far fewer instrument limitations to contend with, and, with the availability of charts, guidelines, practical examples, case histories, and analyses of circuit malfunctions, can acquire considerable knowledge of circuit action the easy way.

Professional electronic troubleshooters know that time is money and that knowledge is power. Your success in the practice of electronic troubleshooting is limited only by the horizons of your technical know-how. The unique, practical application techniques, new methods, trouble-symptom analyses, case histories, troubleshooting guidelines, and technical reference data in this practical book provide key stepping stones to your goal.

Robert G. Middleton

CONTENTS

---

*Indicates a new troubleshooting technique.

CHAPTER 4: DC Current Tests and Measurements 59

CHAPTER 5: AC Voltage Tests and Measurements 73

CHAPTER 6: AC Current Tests and Measurements 89

CHAPTER 7: Audio Troubleshooting Techniques 101

CHAPTER 8: Radio Troubleshooting Techniques 117

SECTION I

ANALOG TROUBLESHOOTING TECHNIQUES

CHAPTER 1

RESISTANCE TESTS AND MEASUREMENTS

CONTINUITY TESTS • POSITIVE AND NEGATIVE TEMPERATURE COEFFICIENTS • DYNAMIC INTERNAL RESISTANCE • TRANSRESISTANCE MEASUREMENT • DIODE TESTS AND MEASUREMENTS • BIPOLAR TRANSISTOR TESTS AND MEASUREMENTS • FINGER TEST • RESISTIVE TOLERANCES • INTERMITTENT RESISTANCE—THE TROUBLESHOOTER'S CURSE

CONTINUITY TESTS

Continuity tests are made to determine whether a continuous path for current flow is present between two points, or whether the path is broken. A continuous path is defined as a low-resistance path because there is no such thing as a zero-resistance path ("dead short") between two test points. A broken path is defined as a high-resistance path because there is no such thing as infinite insulation resistance. Therefore, the terms *short-circuit* and *open-circuit* are comparative resistance definitions.

Continuity tests are usually made with an ohmmeter. Some ohmmeters, such as the multimeter depicted in Figure 1-1, provide a continuous audible tone, in addition to resistance readout, for indication of circuit continuity. This feature permits the trouble-shooter to check continuity without looking back and forth between the tests points and the ohmmeter display.

Although continuity tone tests provide considerable operating convenience, they can also place pitfalls in the path of the unwary electronic troubleshooter. For example, the digital multimeter depicted in Figure 1-1 provides a continuous audible tone when the resistance value between the test points is 179 ohms or less. Consequently, a tape recorder head with a winding resistance of 380 ohms would check "open" on a tone test, although the coil winding is continuous. On the other hand, a cold-solder joint with a resistance of 150 ohms in a power transformer primary circuit would check

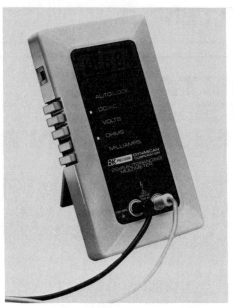

Courtesy, B&K Precision Dynascan Corp

Figure 1-1   A minicomputer-controlled autoranging full-
function digital multimeter.

"continuous" on a tone test, although the connection is "open" for all
practical purposes. *Therefore, it is essential that the troubleshooter be
alert for this kind of pitfall.*

Of course, a low-voltage "open" can be a high-voltage "short,"
as in the case of a high-voltage filter capacitor that arcs internally
when rated working voltage is applied, but which checks "open"
when tested with an ohmmeter. **Moral:** A resistance measurement is
not necessarily conclusive unless it is made at the rated working
voltage of the component or device under test.

POSITIVE AND NEGATIVE TEMPERATURE COEFFICIENTS

Temperature coefficient of resistance is occasionally of importance in troubleshooting procedures. As an illustration, a tempera-
ture-sensitive resistor (thermistor) that has a resistance of 2000 ohms at
room temperature may have a resistance of 75 ohms when heated by a
current flow of 35 mA. The thermistor is said to have a *negative
temperature coefficient of resistance.* This means that the resistance

of the device decreases as its temperature increases. Most semiconductor devices have negative temperature coefficients of resistance.

Metals, on the other hand, have a *positive temperature coefficient of resistance.* This means that the resistance of the metal (such as a copper wire) increases as its temperature increases. For example, if the primary winding of a power transformer measures 14 ohms when cold, it will typically measure 21 ohms when hot. The filament of an incandescent lamp that has a resistance of 27 ohms when cold has a typical resistance of 350 ohms when hot.

DYNAMIC INTERNAL RESISTANCE

Dynamic internal resistance is of basic importance in practical troubleshooting procedures. *Dynamic* means "working" or "operating." On the other hand, *static* resistance means "resting" or "nonoperating." Example:

1. The cold resistance of a lamp filament is 27 ohms. This is static resistance. It can be measured with an ohmmeter.
2. The hot resistance of the lamp filament is 350 ohms when connected to a 117-volt line. This is dynamic resistance. It cannot be measured with an ohmmeter.

Consider the "black box" depicted in Figure 1-2. The black box represents any electrical or electronic network, component, or device. "T" represents any test point in the network—the test point is usually above ground potential while the network is operating normally. There is a definite value of resistance to ground from test point T, but this resistance value cannot be measured with an ohmmeter because the test point is above ground potential. We can easily measure this dynamic internal resistance (DIR) value, however, by applying Ohm's law.

Measurement of Dynamic Internal Resistance

With reference to Figure 1-3, consider the measurement of dynamic internal resistance from test point T to ground. Since T is above ground potential, the DIR cannot be measured with an ohmmeter—the voltage at T would damage the ohmmeter. Therefore we measure the DIR value with a dc voltmeter and a test resistor, as shown in Figure 1-4.

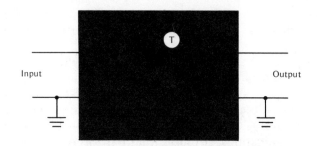

Figure 1-2   The dynamic internal resistance from test point T to ground in the black box cannot be measured with an ohmmeter.

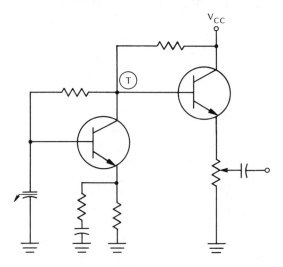

Note: Bipolar transistors are termed current amplifiers (or current-operated devices) because they have a comparatively low input impedance. In turn, a large current change corresponds to a small voltage change. On the other hand, field-effect transistors are termed voltage amplifiers (or voltage-operated devices) because they have a comparatively high input resistance. In turn, a large voltage change corresponds to a small current change. In the strict sense of the terms, there is no such thing as a pure current amplifier, or a pure voltage amplifier. Unless there is at least some voltage change, there can be no current change, and vice versa.

Figure 1-3   The dynamic internal resistance from test point T to ground is measured with a dc voltmeter and a resistor.

Procedure: Connect the DVM between test point T and ground, and measure the dc voltage that is present with the circuit operating normally. Then shunt a resistor R from T to ground, and measure the

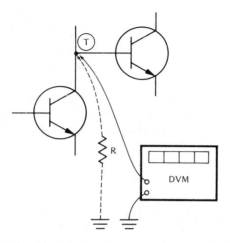

Note: *The value of R should not be too low—excessive current should not be drawn from the test point. Transistors are somewhat nonlinear devices, and if R reduces the voltage at test-point T more than 20 percent, for example, the dynamic internal measurement will have impaired accuracy.*

Figure 1-4   Measurement of dynamic internal resistance from T to ground.

resulting reduced value of dc voltage. Finally, use your pocket calculator to apply Ohm's law and calculate the value of the dynamic internal resistance.

Example: The dc voltage at T measures 5.5 volts with the circuit operating normally. When a 0.5 megohm resistor is shunted from T to ground, the dc voltage decreases to 5.0 volts. This is a change of 0.5 volt. The current drawn by the test resistor is equal to $5/500,000$, or 10 microamperes. In turn, the dynamic internal resistance is equal to $0.5/10 \times 10^6$, or 50,000 ohms.

To briefly recapitulate, resistance is equal to voltage divided by current, and current is equal to voltage divided by resistance. Internal resistance is equal to the change in voltage produced by a shunt load, divided by the current demand of the shunt load.

The dynamic internal resistance of a circuit or network is its most informative resistance value, because it takes into account all of the current paths associated with the test point. Consequently, any malfunction in a component or device included in one of these current paths will result in an abnormal or a subnormal measured value of dynamic internal resistance.

When measuring dynamic internal resistance, it is good practice to choose a shunt resistor that will produce a voltage reduction of approximately 10 percent, and not more than 20 percent, at the test point. In other words, a 10 percent voltage change is within normal range of circuit operation, and permits an accurate DIR measurement. It may be advantageous to use a potentiometer, adjusting its value experimentally as required.

It is evident that dynamic internal resistance can be measured either with signal present or without. In some situations the value of dynamic internal resistance changes when there is signal flow in the circuit. Thereby, a double check of circuit action is available. This double check can be helpful in some troubleshooting situations, which will be explained subsequently.

TRANSRESISTANCE MEASUREMENT

Transresistance measurements can often speed up and simplify troubleshooting procedures in solid-state circuitry; improved accuracy is also provided in low-level amplifier tests. Transresistance is equal to the ratio of output voltage change to input current change for a transistor, circuit, amplifier, or system. For example, if an input current of 10 μA produces an output voltage change of 1 volt in an amplifier stage, the transresistance of the stage is equal to $\frac{1}{10^{-5}}$, or 100 kilohms.

Transistors are basically current-operated devices. This means that a bipolar transistor has comparatively low input resistance. In turn, a small change of input voltage produces a large change of input current. Small changes of input voltage are difficult to measure with service-type voltmeters. On the other hand, it is a simple procedure to inject a precise amount of current (constant current) into the input circuit. The practical result is that the transresistance of a low-level amplifier can be accurately measured, whereas its voltage gain may not be measurable, or may not be measurable with reasonable accuracy.

If a stage has high voltage gain, it will have a high transresistance value. A typical stage with a voltage gain of 61 times has a corresponding transresistance of 92 kilohms.

Constant-Current Source

The troubleshooter needs to use a constant-current source when measuring transresistance so that the value of injected current is known and does not require measurement. With reference to Figure 1-5, a practical constant-current source consists of a 2.25-megohm resistor connected in series with a miniature 22.5-volt battery for injection of 10 μA into an amplifier or network.

DIODE TESTS AND MEASUREMENTS

Silicon and germanium diodes are widely used in electronic equipment. Diodes are commonly checked for front-to-back ratios with an ohmmeter. When the test leads are polarized in one direction, a comparatively low resistance value (forward resistance) is normally indicated. On the other hand, when the test leads are reversed, a very high resistance (reverse resistance) is normally indicated.
Note: The color coding of ohmmeter test leads can be deceptive. Some ohmmeters are designed to apply a positive test voltage via the red lead, and to apply a negative test voltage via the black lead. On the other hand, many ohmmeters are designed to apply a negative test voltage via the red lead, and to apply a positive test voltage via the

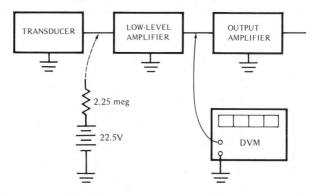

Note: *This is a practical constant-current source because 22.5 volts is large in comparison to the bias voltage in the low-level amplifier. Most transducers, such as tape-recorder heads, are capacitively coupled to the low-level amplifier. In turn, all of the injected constant current flows into the low-level input transistor. Note, however, that the transducer is occasionally direct-coupled to the low-level amplifier. If direct coupling is employed, the injected constant current will branch in proportion to the branch conductances. (Check the schematic diagram to determine if direct coupling might need to be taken into account.)*

Note: *This is a transresistance measurement; transresistance is the ratio of an output voltage change to a corresponding input current change. On the other hand, a transconductance measurement is the ratio of an output current change to a corresponding input voltage change. From a practical viewpoint, transresistance is the parameter of choice for current-operated devices, and transconductance is the parameter of choice for voltage-operated devices.*

Figure 1-5   A high-value resistor and a battery provide a
practical constant-current source.

black lead. Accordingly, the troubleshooter must keep this lack of standardization in mind. Ohmmeter test-lead polarity can be checked with a dc voltmeter.

A silicon diode in normal working condition has an extremely high reverse resistance, and it is virtually unmeasurable with service-type ohmmeters. On the other hand, a germanium diode in normal working condition has a reverse resistance that is detectable, if not accurately measurable, by a high-sensitivity VOM. The reverse resistance is easily measurable with a DVM, provided that it has ranges for megohm values.

Forward-Resistance Values Indicated by Various Types of Ohmmeters

A silicon diode starts to conduct in the forward direction at approximately 0.5 volt; a germanium diode starts to conduct in the forward direction at about 0.2 volt. (See Figure 1-6.) A silicon diode has

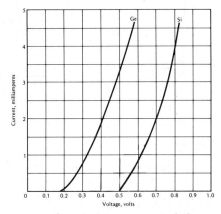

Note: *If the Si voltage-current characteristic were extended several volts in the reverse direction, the current flow would be practically zero, until the Zener point was reached. When the Zener voltage is reached, the reverse current increases with great rapidity. This Zener current can confuse ohmmeter measurements of reverse resistance in silicon diode and transistor devices if the ohmmeter applies sufficient test voltage to reach the Zener point. For example, the Radio Shack 50,000 ohms/volt multitester applies a maximum of 9 volts on its Rx10k range, and applies a maximum of 1.5 volts on its Rx1k range. In turn, a silicon diode or silicon transistor may appear to have low reverse resistance when tested on the Rx10k range of the ohmmeter, whereas the same diode or transistor measures infinite reverse resistance when tested on the Rx1k range of the ohmmeter.*

Figure 1-6 Voltage/current characteristics of typical small-signal silicon and germanium diodes.

higher forward resistance than a germanium diode, and this difference is usually evident in ohmmeter tests. However, since diode forward resistance is nonlinear, any diode will measure different values of forward resistance on various ranges of an ohmmeter. Nevertheless, a troubleshooter who is familiar with a particular type of ohmmeter can distinguish between silicon and germanium diodes on the basis of forward-resistance measurements.

Note the forward-resistance values for a typical silicon diode and for a typical germanium diode on five different types of ohmmeters:

Germanium Diode

1000 ohms/volt Multimeter	20k ohms/volt Multimeter	50k ohms/volt Multimeter	6-range DVM	Auto-ranging DVM
450 ohms	Rx1: 18 ohms	Rx1: 11 ohms	2k: 379 ohms	10.5k
	Rx10: 90 ohms	Rx10: 47 ohms	20k: 10.5k	
	Rx1k: 310 ohms	Rx100: 250 ohms	200k: 10.5k	
		Rx1k: 1600 ohms	2000k: 115k	
		Rx10k: 2000 ohms	20 meg: 150k	

Silicon Diode

1250 ohms	Rx1: 25 ohms	Rx1: 12 ohms	2k: 910 ohms	328k
	Rx10: 200 ohms	Rx10: 88 ohms	20k: —	
	Rx1k: 1100 ohms	Rx100: 675 ohms	200k: 50.8k	
		Rx1k: 5000 ohms	2000k: —	
		Rx10k: 6000 ohms	20 meg: 2.3 meg.	

In general, a silicon diode will measure a higher value of forward resistance than will a germanium diode (on the same range of the same ohmmeter). Note that the first range (200 ohms) of the 6-range DVM was not usable in this example because it was out of range. We also observe that no readout was obtained on the 20k and 2000k ranges of the 6-range DVM; this is because these two ranges are *lo-pwr ohms* ranges. These lo-pwr ohms ranges apply a maximum potential of 0.5 volt to the diode under test. In turn, a silicon diode will not conduct on these lo-pwr ohms ranges. Germanium diodes, on the other hand, *do* show forward conduction on these lo-pwr ohms ranges, because the test voltage is greater than 0.2 volt.

The troubleshooter should keep in mind that some lo-pwr ohmmeters are designed to apply a maximum potential of less than 0.2 volt to the diode under test. If this type of ohmmeter is used, no forward conduction will be obtained for either germanium or silicon diodes on the lo-pwr ohms ranges of the meter (unless the diode is defective). (See Figure 1-7.)

Diodes may become open-circuited or short-circuited. A defective diode often develops leakage (a low value of reverse resistance). In some cases a defective diode will develop both low reverse resistance and high forward resistance.

Making Bad Diodes Out of Good Ones

Troubleshooters should keep in mind the limited forward-current ratings of some mini and micro diode types. In other words, if an ohmmeter produces an excessive forward-current flow through a small diode, the junction will burn out. As an illustration, a 1N34A germanium diode is rated for a steady forward-current flow of 50 mA. This is a conservative rating, and the diode will withstand somewhat greater current flow without burnout. On the other hand, it is poor practice to test the diode at a current flow of 100 mA, for example.

Virtually all ohmmeters employ a 1.5V battery on the Rx1 range. The maximum test current that will flow on the Rx1 range depends on the center-scale indication of the ohms scale. For example, one ohmmeter indicates 10 ohms center scale; another ohmmeter indicates 25 ohms center scale. The maximum current that will flow on the Rx1 range is given by Ohm's law: *An ohmmeter with 10 ohms center-scale indication provides a maximum current flow of $^{1.5}\!/_{10}$, or 150 milliamperes. An ohmmeter with 25 ohms center-scale indication provides a maximum current flow of $^{1.5}\!/_{25}$, or 60 milliamperes.*

Therefore the troubleshooter would check the forward resistance of a 1N34A germanium diode with the ohmmeter that indicates 25 ohms center scale. A brief test at 60 mA forward-current flow is permissible. The procedure of choice with either ohmmeter is to avoid the Rx1 range, and to use the Rx10 range in forward-resistance measurements when the rated forward current is 50 mA or less.

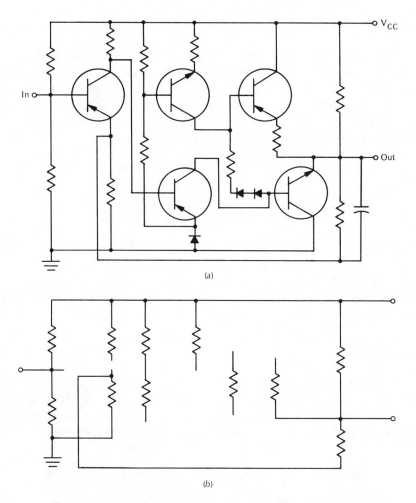

Figure 1-7 In-circuit resistance measurements with lo-pwr ohmmeter. (a) Complete amplifier circuit; (b) lo-pwr ohmmeter "sees" this modified circuit.

Application of excessive heat will also make bad diodes out of good ones. Note that if a germanium diode is warmed between your fingers while measuring its resistance with an ohmmeter, the indicated resistance value will gradually decrease. If the tip of a soldering gun is touched to a diode lead, the indicated resistance will drop quickly. *A heat sink should always be used when diode leads are soldered.* (Thermal damage may cause low reverse resistance, or the diode junction may be destroyed.)

Matched Diodes

Some circuit applications require a pair of matched diodes for normal operation. In other words, both diodes should have the same forward-conduction characteristic. A three-point check of forward-resistance values on three ohmmeter ranges provides a meaningful match test. For example, consider the following forward-resistance measurements on a pair of silicon diodes:

First Diode	Second Diode
Rx1: 12 ohms	Rx1: 10.8 ohms
Rx10: 88 ohms	Rx10: 88 ohms
Rx100: 675 ohms	Rx100: 670 ohms

These test results indicate that this pair of diodes is well matched on the Rx10 and Rx100 ranges of the ohmmeter. On the other hand, they have a 10 percent mismatch on the Rx1 range. Therefore the troubleshooter should select a better matched pair of diodes for use in a critical application.

LED Tests

A light emitting diode (LED) normally has a forward resistance of approximately 20 kilohms and an infinite back resistance when checked with a 20,000 ohms/volt meter. The LED will glow dimly on the Rx1 and Rx10 ranges. A DVM is unsuitable for LED tests.

BIPOLAR TRANSISTOR TESTS AND MEASUREMENTS

An ohmmeter will show whether a transistor is a PNP or an NPN type, whether it is a silicon or germanium type, and will identify its base, emitter, and collector terminals. With reference to Figure 1-8, the transistor is checked out as follows:

1. Measure the resistance between each pair of transistor terminals.
2. The two lowest resistance values are from base to emitter and from base to collector, thereby identifying the base terminal.
3. Whether the transistor is a PNP or an NPN type is shown by the polarity of the ohmmeter test leads in measurement of forward resistance.

4. Whether the transistor is a silicon or a germanium type is shown by the value of forward resistance, based on the troubleshooter's experience with his ohmmeter.
5. The collector and emitter terminals can be identified from the rule that a lower resistance is measured between these terminals when the test voltage is applied in normal operating polarity.

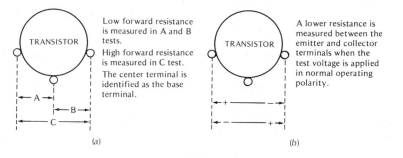

(a)   (b)

Note: *Transistor basing requires attention. When a transistor is replaced, the basing may be different from that of the original transistor, although their appearances are identical.* **Case History:** *An NPN replacement transistor checked out with its collector and emitter terminals reversed, as compared with the original transistor. Moreover, the basing diagram on the replacement transistor packet was incorrect. Therefore, the troubleshooter should not assume that the basing of a transistor is the same as would be expected. Always verify the basing with ohmmeter tests.*

Figure 1-8   Transistor checkout with an ohmmeter. (a) The base terminal has a low forward resistance to each of the two other terminals; (b) a lower resistance is measured between the emitter and collector terminals when the ohmmeter applies a voltage that is polarized as in normal operation. (See also the procedures for a finger test in this chapter.)

FINGER TEST

Unless the ohmmeter has megohm ranges, a finger test must be used to carry out Step 5. In other words, many ohmmeters cannot indicate the very high resistance between the collector and emitter terminals of a silicon transistor. However, a finger test may be used to provide resistance indication, no matter what kind of ohmmeter is used. To make a finger test, the troubleshooter proceeds as follows:

1. Apply the ohmmeter test leads to the collector and emitter terminals of the transistor (which is collector and which is emitter is unknown at this time).
2. Pinch the base lead and one of the other leads between the thumb and forefinger to provide "bleeder resistance." Note the resistance reading, if any.
3. Pinch the base lead and the remaining other lead between the thumb and forefinger to provide "bleeder resistance." Note the resistance reading, if any.
4. Reverse the ohmmeter test leads and repeat Steps 2 and 3.
5. The collector is the terminal that provides the lowest resistance reading when its test voltage is "bled" into the base terminal.

Trick of the trade: If your skin is very dry, and you are using a 1000 ohms/volt meter, moisten your fingers slightly to "bleed" sufficient voltage into the base terminal.

RESISTIVE TOLERANCES

The majority of resistive tolerances in consumer electronic equipment are comparatively "loose." For example, many ±10 percent resistors and quite a few ±20 percent resistors are used. A gold band denotes ±5 percent tolerance; a silver band denotes ±10 percent tolerance; absence of the band denotes a tolerance of ±20 percent. A few resistive tolerances in consumer electronic equipment are comparatively "tight." As an illustration, the resistors in automatic frequency control circuits have a typical tolerance of ±1 percent.

Case history: A VOM became seriously inaccurate on its Rx1 range. For example, a scale reading of approximately 200 ohms was obtained when a 100-ohm resistor was tested. Visual inspection of the ohmmeter circuitry showed that a 9.5-ohm resistor was discolored and had evidently been overheated. The resistor was disconnected for checking, and its actual value was found to be about 8 ohms. When it was replaced with a selected tight-tolerance 9.5-ohm resistor, the ohmmeter was restored to normal operation. (This type of ohmmeter malfunction is commonly caused by applying the ohmmeter test leads in "live" circuitry.)

INTERMITTENT RESISTANCE—THE TROUBLESHOOTER'S CURSE

Intermittents are unquestionably the most difficult kind of circuit fault to pinpoint—an intermittent comes and goes at longer or shorter intervals. Sometimes merely applying voltmeter test leads in a circuit will "cure" an intermittent—temporarily. Most intermittents are unpredictable; the troubleshooter does not know when the intermittent symptom will appear, nor how long it will persist.

The three chief types of intermittents are thermal, mechanical, and electrical. A thermal intermittent responds to temperature variation; a mechanical intermittent is associated with vibration, stress, or movement; an electrical intermittent responds to voltage variations, such as line-voltage fluctuation.

Aggravation of Intermittent Resistance

Troubleshooters use specialized methods to speed up the occurrence of an intermittent condition. This is called aggravation of the intermittent. For example:

1. A thermal intermittent can often be aggravated by heating suspected components or devices with a blast of hot air from a hair dryer. It is often helpful to follow up by cooling the suspected area with a rapidly evaporating liquid spray.
2. Mechanical intermittents can often be aggravated by jarring the assembly, by tapping individual components and devices, or by pulling and twisting flexible leads. Devices and plugs should be moved up and down in their sockets.
3. An electrical intermittent can sometimes be aggravated by switching the equipment on and off several times in succession. Some electrical intermittents can be initiated by operating the equipment at somewhat abnormal or subnormal supply voltage.

Case history: The picture rolled intermittently on the screen of a TV receiver, requiring occasional resetting of the vertical-hold control. When the troubleshooter adjusted the control, he noted that it did not seem to be as "smooth" as could be desired. He checked the potentiometer with an ohmmeter and observed that the scale indication was not completely stable—the pointer fluctuated slightly and erratically, particularly when the potentiometer was tapped. When the potentiometer was replaced, the receiver was restored to normal operation.

CHAPTER 2

DC VOLTAGE
TESTS AND
MEASUREMENTS

CIRCUIT LOADING BY DC VOLTMETERS • CIRCUIT DETUNING BY DC VOLTMETERS • WHEN TWO ISOLATION RESISTORS ARE NEEDED • DC VOLTAGE TOLERANCES • DC VOLTMETER RESPONSE TO DC PULSES • PRACTICAL EXAMPLES • PEAK HOLD UNIT FOR DC VOLTMETER • MEASUREMENT OF DC PULSE WIDTH • PRACTICAL EXAMPLES • COPING WITH NARROW PULSES • INTERMITTENT VOLTAGE—A TROUBLESHOOTER'S NIGHTMARE • EXPERIMENTAL PROJECTS

CIRCUIT LOADING BY DC VOLTMETERS

When a VOM, TVM, or DVM is applied in a circuit to measure the dc voltage at a selected point, the instrument draws more or less dc current from the circuit; in turn, the voltmeter is said to load the circuit. Excessive circuit loading is a pitfall that lies in wait for the unwary electronic troubleshooter. Accordingly the basic considerations involved in circuit loading should be kept in mind.

As explained later in this chapter, a slightly elaborated procedure can be used to measure dc voltage without drawing any current from the circuit under test. However the troubleshooter will need to use this elaborated procedure only in special types of tests. For example, zero circuit loading must often be observed in measuring the rise time of a square wave, and in measuring the amplitude of digital pulses with low repetition rates.

The loading that a voltmeter imposes on a circuit under test is determined by the input resistance of the voltmeter in comparison to the internal resistance of the circuit. A simple example is shown in Figure 2-1. The internal resistance of this circuit is 250 kilohms. A 1000 ohms/volt meter shown in (a) has an input resistance of 15 kilohms on its 15V range. In turn, its input resistance is much less than the internal resistance of the circuit under test. The VOM indicates 0.51 volt, and the measurement error is 94 percent.

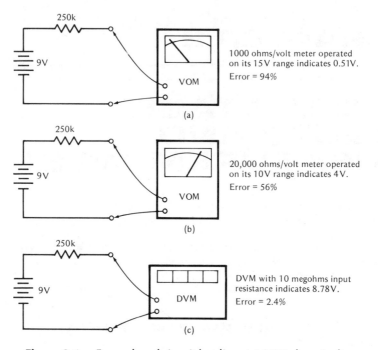

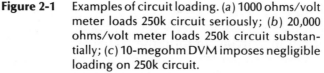

Figure 2-1   Examples of circuit loading. (a) 1000 ohms/volt meter loads 250k circuit seriously; (b) 20,000 ohms/volt meter loads 250k circuit substantially; (c) 10-megohm DVM imposes negligible loading on 250k circuit.

Next, the 20,000 ohms/volt meter depicted in Figure 2-1 (b) has an input resistance of 200 kilohms on its 10V range. Its input resistance is comparable with the internal resistance of the circuit under test. This VOM indicates 4 volts, and the measurement error is 56 percent.

Finally, the DVM depicted in (c) has an input resistance of 10 megohms on all ranges. Its input resistance is much higher than the internal resistance of the circuit under test. The DVM indicates 8.78 volts, and the measurement error is 2.4 percent.

CIRCUIT DETUNING BY DC VOLTMETERS

Electronic circuits can be detuned, and circuit action disturbed, by the capacitance between the test leads of a voltmeter, although the input resistance to the voltmeter may be very high. A typical DVM has an input capacitance of 25 pF. This is not a large capacitance as far as the audio troubleshooter is concerned, but the FM radio troubleshooter and the TV troubleshooter find that 25 pF can seriously disturb high-frequency circuit action in various situations.

This hazard is most apparent in high-frequency oscillator circuits in which part or all of the base-emitter bias is signal-developed. If the input capacitance of a voltmeter "kills" the oscillator, it also "kills" the dc bias voltage, and thereby makes a "good" oscillator circuit look "bad." Therefore many TVM and DVM manufacturers provide isolation probes for their voltmeters. An isolation probe usually consists of a 100-kilohm resistor in series with the "hot" lead to the meter. The isolation probe is used only when troubleshooting high-frequency circuitry.

Of course a 100-kilohm resistor connected in series with the hot lead to the voltmeter reduces the indicated voltage; to obtain the true dc voltage value, the troubleshooter multiplies the indicated voltage value by 1.01. (Since the measurement error is only 1 percent, it is generally ignored in practical procedures.)

If a 20,000 ohms/volt meter or a 50,000 ohms/volt meter is used instead of a TVM or DVM, a 100-kilohm isolation resistor is similarly required to avoid "killing" high-frequency oscillators. Note that the scale correction factor depends on the range that is being used; if the 10V range of a 20,000 ohms/volt meter is being used, the scale correction factor is 1.5 times the indicated voltage value. Or, if the 5V range of a 50,000 ohms/volt meter is being used, the scale correction factor is 1.4 times the indicated voltage value.

WHEN TWO ISOLATION RESISTORS ARE NEEDED

As shown in Figure 2-2, the troubleshooter will occasionally encounter oscillatory circuits in which all three terminals of the transistor operate above rf ground potential. In this situation the base-emitter bias voltage cannot be measured directly by connecting one test lead to the base terminal and connecting the other test lead to the emitter terminal. Circuit action is likely to be seriously disturbed due to circuit detuning by the nonisolated test lead. Accordingly, the troubleshooter employs either of the following procedures:

1. The base voltage is measured with respect to V_{CC} or to ground, and the emitter voltage is measured with respect to V_{CC} or to ground. The difference between these two readings is the base-emitter bias voltage.

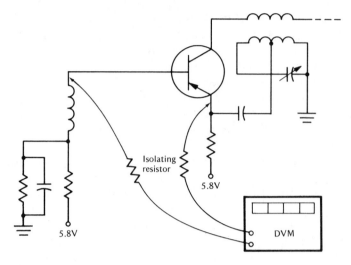

Note: *A 100-kilohm isolating resistor in series with each of the DVM test leads ensures that minimum disturbance will be imposed on the high-frequency oscillator circuit.*

Figure 2-2 Transistor oscillates with all three terminals operating above rf ground potential.

2. The base-emitter bias voltage is measured directly by connecting a 100-kilohm isolation resistor in series with the "hot" lead of the voltmeter, and also connecting a 100-kilohm isolation resistor in series with the return lead of the voltmeter. (The scale correction factor is doubled in this situation.)

DC VOLTAGE TOLERANCES

The electronic troubleshooter must evaluate dc voltage values that have comparatively "loose" tolerances, and others that have comparatively "tight" tolerances. As an illustration, the collector voltage in an amplifier stage is noncritical, whereas the base voltage is quite critical as indicated in Figure 2-3. With a typical transistor, stage operation is optimum when R has a value of 200 kilohms, and the base voltage is 0.65 volt. On the other hand, when R has a value of 50 kilohms and the base voltage is 0.68 volt, the transistor is near saturation and the stage overloads at a low input voltage. Thus, the base-voltage tolerance is "tight" in the foregoing example—on the order of 1 or 2 percent. If the base voltage varies 5 percent, the stage is practically unworkable. On the other hand, the collector voltage has a

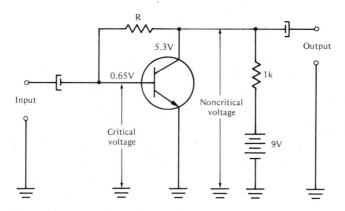

Figure 2-3   Example of critical and noncritical dc voltages.

"loose" tolerance—it can vary 20 percent without seriously affecting stage operation.

DC VOLTMETER RESPONSE TO DC PULSES

Electronic troubleshooting procedures are concerned with pulse tests and measurements in various situations. For example, a dc voltmeter is used to measure the average value of a dc pulse train, and to measure the peak value of the dc pulse train. With reference to Figure 2-4, dc pulses have a positive excursion but no negative excursion (or a negative excursion but no positive excursion). If a dc pulse train is applied to a VOM, TVM, or DVM, the dc function of the voltmeter indicates the average value of the pulse train.

The average value of a pulse waveform denotes the voltage level at which excursions above and below this level have equal areas (equal amounts of electric charge). The average value of the dc pulse waveform is important because *it tells the troubleshooter how wide the pulses are with respect to the pulse period.*

As indicated in Figure 2-4, the ratio of pulse peak voltage to pulse average voltage is equal to the ratio of pulse period to pulse width.

PRACTICAL EXAMPLES

1. A dc pulse train measures 0.5 volt on the dc-voltage function of a DVM. This is the average value of the pulse waveform.
2. The dc pulse train measures 1.0 volt on the peak-hold function of the DVM. This is the peak value of the pulse waveform.

3. **Conclusion:** Since the average voltage is 0.5 of the peak voltage, the voltmeter is "looking at" a square wave (a dc square waveform).

4. Another dc pulse train measures 0.1 volt on the dc-voltage function of a DVM. This is the average value of the pulse waveform.

5. The dc pulse train measures 1.0 volt on the peak-hold function of the DVM. This is the peak value of the pulse waveform.

6. **Conclusion:** Since the average voltage is 0.1 of the peak voltage, the voltmeter is "looking at" a pulse train in which the pulse width is 0.1 of the pulse period.

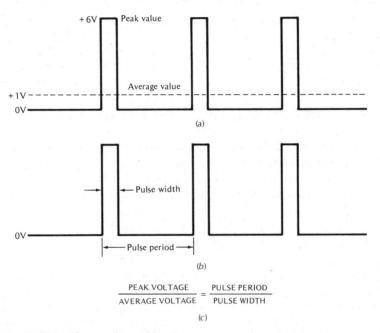

$$\frac{\text{PEAK VOLTAGE}}{\text{AVERAGE VOLTAGE}} = \frac{\text{PULSE PERIOD}}{\text{PULSE WIDTH}}$$

(c)

Use CAT! Punch out pulse-width values on your pocket calculator—it can speed up and simplify pulse-width calculations.

Note: *The shortest pulse period and the narrowest pulse width that you can analyze with your dc voltmeter depends upon its frequency capability. Use a pulse generator to check out the capability of your particular meter. As the upper frequency capability of the meter is approached, the scale reading will decrease for one brand of meter, and will increase for another brand of meter, although the input peak voltage and the pulse-period/pulse-width value is maintained constant.*

Figure 2-4 Basic dc pulse relations. (a) Peak value and average value; (b) pulse period and pulse width; (c) relation of peak voltage and average voltage to pulse period and pulse width.

PEAK-HOLD UNIT FOR DC VOLTMETER

Many DVMs do not have a peak-hold function. In that case, the troubleshooter uses a peak-hold unit to measure the peak voltage of a dc pulse train. As illustrated in Figure 2-5, a positive peak-hold unit's diode is oppositely polarized to that in a negative peak-hold unit. Note that a peak-hold unit cannot be used with a VOM to measure the peak voltage of a dc pulse train. In other words, a VOM draws appreciable current from the peak-hold unit, with the result that the measurement is in substantial error. On the other hand, a peak-hold unit can be used with a TVM to measure the peak voltage of a dc pulse train. (Like a DVM, a TVM has a very high input resistance, such as 10 megohms.)

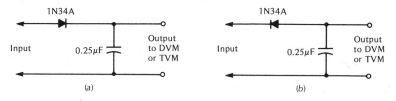

Note: The positive peak-hold unit is used to measure the peak voltage of a positive dc pulse train. The negative peak-hold unit is used to measure the peak voltage of a negative dc pulse train.

To avoid damage to the 1N34A diode, the peak input signal voltage should not exceed 75 volts.

Note: A germanium diode has a barrier potential of approximately 0.2 volt. From a practical viewpoint, this means that the DVM indicates a voltage value that is 0.2 volt less than the input voltage.

Figure 2-5   Peak-hold units for use with a DVM or TVM. (a) Positive peak-hold configuration; (b) negative peak-hold configuration.

MEASUREMENT OF DC PULSE WIDTH

Measurement of dc pulse width is based on the relations given in Figure 2-4(c). In other words, we can measure the pulse width in milliseconds or microseconds if we know the peak voltage, the average voltage, and the pulse period in milliseconds or microseconds. To measure the pulse period, proceed as follows:

1. As exemplified in Figure 2-6, mix the output from the pulse source with the output from a sine-wave source, such as an audio oscillator.

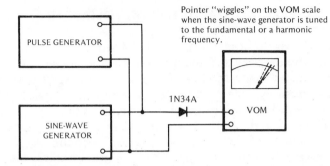

Note: *Pulse tests and measurements made with meters are not as accurate as if an oscilloscope is used. However if a troubleshooter does not have an oscilloscope available, the meter technique can "pinch hit."*

Note: *The 1N34A diode must be polarized to match the polarity of the dc pulse train.*

Operate the VOM on its dc voltage function, and on a range that provides at least half-scale deflection.

The largest zero-beat indication is obtained when the sine-wave generator output voltage is adjusted to equal the pulse-generator output voltage.

Figure 2-6   Simple example of pulse-width measurement.

2. Feed the mixed signal through a diode to a VOM or TVM.
3. As the sine-wave generator is tuned near to the frequency corresponding to the pulse period, the pointer will "wiggle" on the scale of the dc voltmeter.
4. Tune the sine-wave generator for zero beat (halfway through the wiggle interval).
5. The sine-wave generator frequency is then equal to the fundamental frequency of the pulse train.
6. The pulse period is equal to the reciprocal of this fundamental frequency. For example, if the fundamental frequency is 1 kHz, the pulse period is 1 millisecond.

Note that various wiggle intervals will occur as the sine-wave generator frequency is increased. However the largest wiggle occurs when the sine-wave generator is tuned to the fundamental frequency of the pulse train. The smaller wiggles at higher sine-wave frequencies indicate harmonic frequencies in the pulse train.

PRACTICAL EXAMPLES

1. A dc pulse train measures 0.5 volt on the dc-voltage function of a DVM. This is the average value of the dc pulse waveform.

2. This dc pulse train measures 1.0 volt on the peak-hold function of the DVM. The troubleshooter then knows that the DVM is "looking at" a square waveform.
3. With reference to Figure 2-6, the fundamental frequency of the dc pulse train is indicated as 1 kHz by the sine-wave generator. (A 20,000 ohms/volt VOM was operated on its 0.25-volt range in this example.)
4. Since the fundamental frequency of the pulse train is 1 kHz, the pulse period is 1 millisecond.
5. Another dc pulse train measures 1.0 volt peak, and 0.1 volt average. In this case the troubleshooter knows that the DVM is "looking at" a pulse train in which the pulse width is equal to 0.1 of the pulse period.
6. The foregoing pulse period measures 1 millisecond. Therefore, the pulse width is equal to 0.1 millisecond.

COPING WITH NARROW PULSES

A peak-hold unit such as that shown in Figure 2-5 will not provide accurate peak-voltage measurement of very narrow pulses. In other words, even a DVM with 10-megohm input resistance draws a small amount of current from the peak-hold unit, and very narrow pulses do not have sufficient energy to supply this small current demand of the DVM. In turn, the DVM indicates a subnormal voltage value.

To obtain accurate peak-voltage measurements of very narrow pulses, a reverse-biased peak-hold unit may be used, as shown in Figure 2-7. Observe that the bias-box voltage opposes current flow from the peak-hold unit into the DVM. The peak-voltage measurement procedure is as follows:

1. Apply the dc pulse waveform to the peak-hold unit input.
2. Adjust the bias-box voltage to a value such that the polarity indication of the DVM fluctuates between + and −. (The DVM is being operated as a sensitive null indicator.)
3. The meter on the bias box now indicates the peak voltage of the dc pulse train.

Note that a reverse-biased peak-hold unit effectively works into an infinite impedance (resistance), and in turn zero current is drawn by the DVM. Nevertheless, high accuracy requires that the hold capacitor have very high insulation resistance, and that the hold diode have very high reverse (back) resistance. Accordingly, it is good practice to use a selected capacitor and a selected diode in the peak-hold unit.

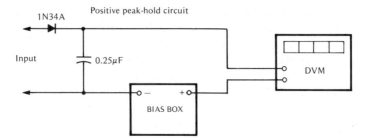

Note: *The bias-box voltage is polarized to oppose current flow into the DVM. If a negative peak-hold unit is used, the bias-box polarity is reversed. As noted previously, the barrier potential of a germanium diode is approximately 0.2 volt. From a practical viewpoint, this means that the bias-box voltage will be 0.2 volt less than the input voltage when the DVM indicates zero (null).*

Figure 2-7   Reverse-biased load circuit for positive peak-
hold unit used to measure the peak voltage of
very narrow dc pulses.

INTERMITTENT VOLTAGE—A TROUBLESHOOTER'S NIGHTMARE

Intermittent voltage symptoms are comparatively difficult to troubleshoot because the fault condition appears and disappears erratically. To close in on the malfunction, it is often helpful to monitor equipment operation with two or more voltmeters connected at key test points in the circuitry. Then, when the trouble symptom occurs, the voltmeter with the incorrect reading serves to localize the malfunctioning area. Aggravation of intermittents was explained in Chapter 1.

Case history: The audio output from both channels of a Realistic SA-10 stereo amplifier became intermittently weak and distorted. The intermittent condition could be aggravated by tapping the PC board, particularly at the input end. Then the dc voltages at the input and driver transistors were monitored. When the intermittent occurred, the collector voltages of the input transistors dropped from 4.5V to 0.5V; the collector voltages of the driver transistors remained normal. With reference to Figure 2-8, inspection of the collector circuit for Q101 showed that R106 (which had been previously replaced) had an unsoldered lead in one of the PC board eyelets. When the connection to R106 was properly soldered, the intermittent trouble was corrected.

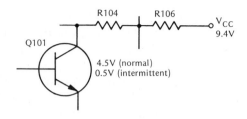

Figure 2-8   R106 made intermittent connection to the PC board.

EXPERIMENTAL PROJECTS

Measurement of Power-Supply Voltage With and Without a Filter Capacitor

This project utilizes a small power transformer and a filter capacitor, with a VOM, TVM, or DVM. The experimental circuit is shown in Figure 2-9. Parts required are:

1. Doorbell or doorchimes transformer with 6 to 10 volts output
2. Power cord with plug
3. Electrolytic capacitor, 1000 μF
4. Silicon rectifier diode, 2.5 ampere rating
5. Resistor, 20 ohms, 5 watts rating

First connect the rectifier diode in series with the secondary of the transformer, as shown in Figure 2-9(a). Plug the primary cord into a 117-volt 60-Hz outlet. Measure the dc voltage between terminals A and B. Note that the red lead of the dc voltmeter is connected to terminal A, and the black lead is connected to terminal B. *If the pointer does not swing up-scale, the diode is connected into the circuit with wrong polarity.*

Next note the voltmeter reading. Then connect the filter capacitor between terminals A and B, observing correct polarity. (Figure 2-9(*b*).)
Caution: If the polarity of the filter capacitor is accidentally reversed, the circuit will seem to operate normally at the start. However, the capacitor will leak excessively and will heat up to the point that it is destroyed.

Observe the voltmeter reading with the filter capacitor connected between terminals A and B. If no error has been made in the experiment, the dc-voltage reading will increase approximately 3.14 times. For example, if the dc-voltage value without a filter capacitor is 2.6 volts, you would expect the dc-voltage value with a filter capacitor to measure about 8.16 volts.

Comment: You are very unlikely to find that the second reading is *exactly* 3.14 times the first reading. Experimental errors include the form factor of the 117-volt waveform (as explained subsequently), the effect of barrier potential in the rectifier diode, accuracy of dc-voltmeter indication, and precision in reading the scale.

Next note the voltmeter reading with the filter capacitor connected between terminals A and B. Then connect the 20-ohm resistor across the capacitor terminals, as shown in Figure 2-9(c).

Observe the voltmeter reading with the resistive load connected between terminals A and B. The dc-voltage reading will normally decrease approximately 20 percent. For example, if the unloaded filter has an output of 8.75 volts, you would expect the voltage to decrease to approximately 7 volts when the load resistor is connected into the circuit.

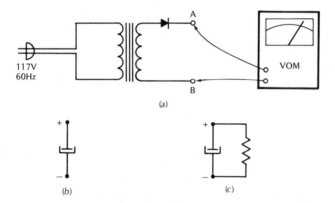

Figure 2-9 Basic power supply. (a) Stepdown transformer and rectifier diode; (b) electrolytic filter capacitor; (c) filter capacitor with resistive load.

Comment: You are not likely to find that the loaded filter output voltage decreases *exactly* 20 percent from its unloaded value. The dc-voltage reading under load is affected by the actual value of the filter capacitor, by the tolerance on the resistor value, by the form factor of the supply voltage, and by the precision of reading the scale.

Next disconnect the capacitor from terminals A and B and note the decrease that results in dc output voltage. Typically the loaded but unfiltered dc output voltage will drop about 67 percent. For example, if loaded and filtered output voltage is 7 volts, disconnection of the filter capacitor causes the output voltage to drop to 2.3 volts (to 33 percent of its initial value).

Comment: You are unlikely to find that the unfiltered dc output voltage is *exactly* equal to 33 percent of its filtered value. The decrease that occurs is affected by the resistance of the transformer secondary winding, the actual value of the filter capacitor, the tolerance on the resistor value, the form factor of the supply voltage, and by the precision of reading the voltmeter scale.

Measurement of Battery Voltage
Under No-Load and Load Conditions

This project utilizes a pocket calculator, a new 9-volt battery, and one or more used batteries. Calculators generally use alkaline batteries. The terminal voltage of a fresh 9-volt battery measures slightly over 9 volts. As the battery ages its terminal voltage progressively decreases.

First connect a new 9-volt battery to the calculator. Measure the battery voltage with the calculator turned off, then measure the battery voltage with the calculator turned on. Next connect a used 9-volt battery to the calculator. Measure the battery voltage with the calculator turned off, then measure the battery voltage with the calculator on. Compare the voltage readings obtained under no-load and full-load conditions for the new and used batteries. Note that when a battery ages, its no-load terminal voltage decreases, and its percentage of voltage decrease under load becomes greater.

Examples: A new 9-volt battery measured 9.2 volts under no load; when the calculator was in operation the battery voltage measured 8.95 volts. This represents a 2.8 percent decrease in terminal voltage under load. Next, a used 9-volt battery measured 9.0 volts under no load; when the calculator was in operation the battery voltage measured 8.5 volts. This represents a 13 percent decrease in terminal voltage under load. Another used 9-volt battery measured 7.85 volts under no load, but when the calculator was in operation the battery voltage measured 7.2 volts. This represents a 15 percent decrease in terminal voltage under load.

Measurement of Voltage "Soaked Up"
by an Electrolytic Capacitor

This project utilizes a large electrolytic capacitor, such as that used in the power supply for an elaborate digital computer. It demonstrates the residual charge retained by the capacitor.

First connect a large electrolytic filter capacitor, such as an 85,000 μF 10-volt capacitor, to a 6-volt battery. Connect the negative terminal of the battery to the negative terminal of the capacitor. Connect the positive terminal of the battery to the positive terminal of the capacitor.

Next disconnect the battery from the capacitor and measure the capacitor terminal voltage with a dc voltmeter.

Then short-circuit the capacitor terminals to discharge the capacitor. Disconnect the short-circuit wire and measure the capacitor terminal voltage.

Let the capacitor stand open-circuited for a minute, then measure the capacitor terminal voltage. Note any increase of terminal voltage that may have occurred while the capacitor was standing open-circuited.

Example: An 85,000 μF 10-volt electrolytic capacitor was charged to 6 volts. The capacitor was then short-circuited, and its terminal voltage measured zero. However, after the capacitor remained open-circuited for a minute, its terminal voltage measured 0.4 volt. In other words, the capacitor had "soaked up" 0.4 volt while it was charged to 6 volts.

CHAPTER 3

DC VOLTMETER PROBES AND CLAMPS

PEAK PROBES • FORWARD-BIASED PEAK-READING PROBE • REVERSE-BIASED PEAK-READING PROBE • POSITIVE-PEAK/NEGATIVE-PEAK/PEAK-TO-PEAK PROBE • TROUBLESHOOTING WITH CLAMPS • CAPACITOR-DIVIDER UNIT FOR HIGH PEAK-VOLTAGE MEASUREMENTS • DEMODULATOR PROBE • DISTORTION TEST PROBE • ONE-SHOT PEAK-HOLD UNIT • ONE-SHOT DIP-HOLD UNIT

PEAK PROBES

Peak-reading probes are extensively used by electronic trouble-shooters to measure positive-peak and negative-peak voltages in complex ac waveforms. Peak-reading (peak-indicating) probes are also called signal-tracing probes. As shown in Figure 3-1, a basic peak-reading probe consists of a capacitor and a diode. The probe measures the peak voltage of the positive excursion or the peak voltage of the negative excursion in an ac waveform.

Although a sine wave has equal positive-peak and negative-peak voltages, a complex ac waveform, such as the sync-pulse train depicted in Figure 3-2, has unequal positive-peak and negative-peak voltages. When the voltage waveform in Figure 3-2 is fed into a positive-peak probe, the DVM will indicate its positive-peak voltage. On the other hand, when the same waveform is fed into a negative-peak probe, the DVM will indicate its negative-peak value.

FORWARD-BIASED PEAK-READING PROBE

When peak-reading probes are applied in *low-level circuits,* a limitation is imposed on indication sensitivity by the diode in the probe, regardless of the DVM sensitivity. In other words, a germanium diode does not conduct until the signal level is at least 0.2 volt; a

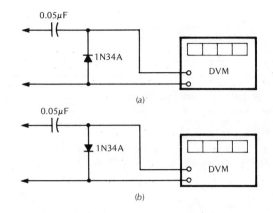

(a)

(b)

Note: These probes have a useful frequency range from approximately 500 Hz to 10 MHz. At frequencies below 500 Hz, the capacitor progressively attenuates the signal voltage. At frequencies above 10 MHz, the input leads to the DVM react as tuned line sections and cause standing-wave errors.

To avoid possible damage to the 1N34A diode, the input signal voltage should not exceed 25 rms volts.

These probes will objectionably load circuits that have a dynamic internal impedance greater than 1500 ohms. (Dynamic internal impedance is explained in Chapter 5.)

Figure 3-1 Peak-reading probes for TVM or DVM. (a) Positive-peak probe; (b) negative-peak probe.

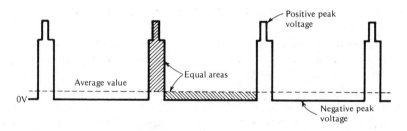

Important: Although the average value of an ac waveform is zero, it will read zero on the dc function of a VOM only if the peak voltage of the waveform does not exceed the range that is being used on the meter. For example, if the positive-peak voltage of the ac waveform is 8 volts, a dc voltmeter will indicate zero on its 10-volt range. On the other hand, the dc volt-meter will not indicate zero on its 2.5-volt range. The reason is that the 8-volt peak in the ac waveform produces dc current flow through the meter's protective rectifier(s).

Figure 3-2 Example of unequal positive-peak and nega-tive-peak voltages.

silicon diode does not conduct until the signal level is at least 0.4 volt. This is just another way of saying that a low-level signal with a peak amplitude of 199 millivolts produces zero output from a basic peak-indicating probe.

If a small forward bias is applied to the diode in a peak-indicating probe, very low signal levels can be traced (although the full peak voltage value is not indicated by the DVM). This small forward bias is applied by means of a bias box connected in series with the DVM, as shown in Figure 3-3. A germanium diode requires approximately 0.2 volt of forward bias; a silicon diode requires approximately 0.4 volt of forward bias. When the bias voltage is adjusted so that the DVM is just beginning to indicate (for example, fluctuating between a zero reading and a 1 mV reading), the probe is then in its condition of maximum sensitivity. In turn, the DVM will respond to a very low-level signal input to the probe. Note that a silicon epoxy diode has very high reverse resistance and will provide high-precision measurements. Note also that if the small forward-bias voltage produces a zero-offset indication on the TVM or DVM (such as 1 or 2 mV), this offset can be subtracted from the readout for maximum precision measurement.

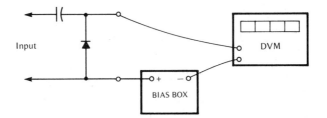

Note: *This is a forward-biased positive-peak probe. If a negative-peak probe is used, the bias-box polarity is reversed.*

The capacitor should be selected for very high insulation resistance, and the diode should be selected for very high reverse resistance.

Caution: *When used with a TVM or DVM, a false readout may be obtained if the ac component of the output from the peak-reading probe exceeds the range being used on the dc voltmeter. (Make a cross-check, using the next higher range on the meter.)*

Point-contact diode: *The old-style point-contact diode differs from the modern junction diode in that the former has zero barrier potential. As such it has practically the same small-signal sensitivity whether it is zero biased or forward biased. However the troubleshooter may find it difficult or impossible to locate a source for point-contact diodes. Therefore small-signal tests must usually be made with forward-biased junction diodes.*

Figure 3-3   Forward bias applied to a peak-reading probe
for maximum sensitivity to low-level signals.

REVERSE-BIASED PEAK-READING PROBE

Electronic troubleshooters sometimes need to measure the peak voltage of an ac pulse waveform that has *very narrow pulses.* In this situation a basic peak-reading probe indicates less than the true peak voltage because very narrow pulses do not have sufficient energy to supply the DVM measuring-current demand. However true peak-voltage values can be measured as follows. With reference to Figure 3-4, the current demand of the DVM is eliminated by connecting a bias box in series with the DVM. The narrow-pulse waveform is applied to the probe input, and the bias box is then adjusted for zero indication on the DVM. That is, the DVM indication will fluctuate from + to − as the readout fluctuates slightly around zero. (The DVM is being operated as a sensitive null indicator.) Now the bias-box meter indicates the true peak value of the applied waveform.

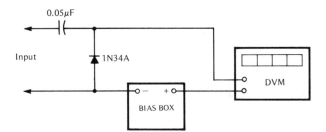

Note: The bias-box voltage is polarized to back-bias the diode in a positive-peak probe. If a negative-peak probe is used, the bias-box polarity should be reversed.

For maximum indication accuracy, the capacitor should have very high insulation resistance, and the diode should have very high reverse resistance.

Note: Since the germanium diode has a barrier potential of approximately 0.2 volt, the bias-box voltage will be 0.2 volt less than the input voltage when the DVM indicates zero (null indication).

Figure 3-4 Reverse-biased peak-reading probe for measuring peak voltage of very narrow pulses.

POSITIVE-PEAK/NEGATIVE-PEAK/PEAK-TO-PEAK PROBE

A combination positive-peak/negative-peak/peak-to-peak probe arrangement is shown in Figure 3-5. This type of probe speeds up troubleshooting procedures because the three principal voltages in a complex ac waveform can be measured with a single probe. On the other hand, the combination probe has the disadvantage that it imposes twice the circuit loading that is imposed by a positive-peak probe or a negative-peak probe.

Conversion of DC Pulses Into AC Pulses

During the course of troubleshooting pulse circuitry the technician occasionally needs to convert a dc pulse train into an ac pulse train—for example to measure peak voltages with basic probes. This is easily accomplished, as shown in Figure 3-6—the technician merely connects a capacitor in series with the output from the dc pulse source. The resulting capacitor output is a corresponding ac pulse waveform.

TROUBLESHOOTING WITH CLAMPS

The opposite requirement also occurs occasionally during the course of troubleshooting pulse circuitry. In other words, the technician needs to convert an ac pulse train into its corresponding dc pulse train—for example, to measure pulse width. This conversion is also easily accomplished by means of a clamp, as shown in Figure 3-7. An ac pulse waveform can thereby be changed into either a positive-pulse waveform or a negative-pulse waveform.

For example, to measure the average value of the pulses in an ac pulse train, we connect a clamp at the output of the ac pulse source. Then we connect a DVM at the output of the clamp. The dc voltage measured at the clamp output is equal to the average value of the dc pulse train. (To measure the peak-to-peak voltage of the ac pulse train, we connect a peak-to-peak probe at the output of the ac pulse source, as previously explained.)

The question often asked is "What is the difference between a peak-reading probe and a clamp to ground?" The difference is solely in the evaluation of output. In the case of a peak-reading probe, we are concerned with the *dc* output from the probe. In the case of a clamp, we are concerned with the *ac* output from the probe. As an illustration, the output may be fed to a dc voltmeter, or the output may be fed to the grid of a color picture tube in a TV receiver. Note that the dc voltmeter "uses" the dc output from the clamp, whereas the grid in the color picture tube "uses" the ac output from the clamp. When a utilization device processes the ac output, we commonly call the clamp a *dc restorer*.

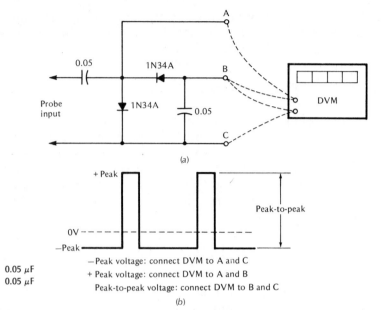

(a)

(b)

−Peak voltage: connect DVM to A and C
+ Peak voltage: connect DVM to A and B
Peak-to-peak voltage: connect DVM to B and C

0.05 μF
0.05 μF

Note: *The useful frequency range of this probe is from about 500 Hz to 10 MHz.*

This probe will objectionably load circuits that have a dynamic internal impedance greater than 700 ohms.

Figure 3-5 Positive-peak/negative-peak/peak-to-peak probe. (a) Configuration; (b) Example of pulse-voltage measurements.

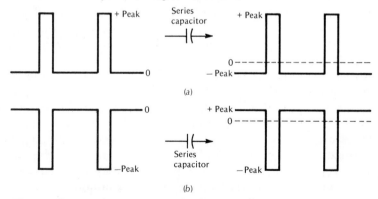

(a)

(b)

Note: *The ac pulse waveform in (a) is said to have* positive-going *pulses.*

The ac pulse waveform in (b) is said to have negative-going *pulses.*

Figure 3-6 A series capacitor changes dc pulses into ac pulses. (a) Positive dc pulses changed into ac pulses; (b) negative dc pulses changed into ac pulses.

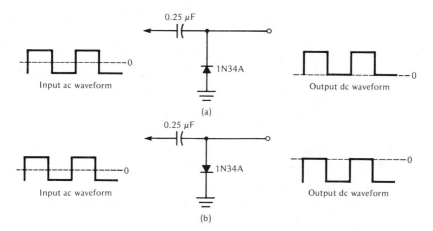

Note that the clamp in (a) converts an ac pulse waveform into a positive-pulse waveform. Conversely, the clamp in (b) converts an ac pulse waveform into a negative-pulse waveform.

Figure 3-7   Clamp circuits for clamping to ground. (a) Negative peaks clamped to ground; (b) positive peaks clamped to ground.

CAPACITOR DIVIDER UNIT FOR HIGH PEAK-VOLTAGE MEASUREMENTS

Electronic troubleshooters occasionally encounter high ac voltages. If the peak voltage does not exceed 1 kilovolt, probes such as those depicted in Figure 3-1 can be used, provided that the capacitor and resistor in the probe circuit are rated for 1-kV operation. Note that the diode must withstand a peak inverse voltage equal to double the capacitor voltage. Accordingly, if 1-kV diodes such as the Archer 276-1114 type are used, two diodes should be connected in series, as shown in Figure 3-8.

Peak voltage measurements in high-voltage ac circuits can also be measured with general-purpose peak-reading probes, elaborated with a capacitor voltage divider unit, as shown in Figure 3-9. In other words, the 11-pF and 1000-pF capacitors function as a capacitor voltage divider and as rectified charge-storage capacitors. To calibrate the arrangement for precise 100-to-1 attenuation, the series resistance R_c is selected as required.

Note that the probes depicted in Figures 3-8 and 3-9 will load circuits that have high dynamic internal resistance. To minimize loading of the circuit under test, a bias box may be employed in series with the DVM, as shown in Figure 3-4.

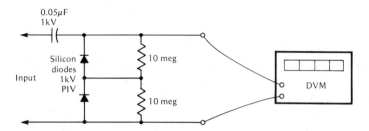

Note: *The 10-meg resistors are used to equalize the peak-inverse voltages (PIV) across the diodes, and thereby avoid possible breakdown.*

Figure 3-8 A 1-kV peak-reading probe arrangement.

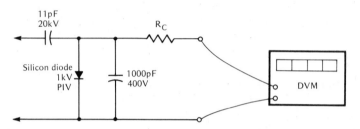

Note: R_c *is a calibrating resistor; its required value depends on the tolerances of the probe components. (In a typical arrangement, a 100-to-1 input/output ratio required a 5.6 megohm calibrating resistor.)*

Figure 3-9 A 100-to-1 capacitance-divider peak-reading probe configuration.

DEMODULATOR PROBE

Demodulator probes are used to signal-trace in rf circuits, and to make amplitude-modulation checks. A demodulator probe, such as that shown in Figure 3-10, is also called a detector probe or a traveling detector. When an rf carrier voltage is applied to the probe, a corresponding dc voltage is applied to the DVM. When an amplitude-modulated rf signal is applied to the probe, the output from the probe consists of a dc voltage corresponding to the carrier level and also an ac audio waveform (the modulating signal).

When the demodulator probe is used with a DVM that has the same input resistance on its dc-voltage and ac-voltage functions, a percentage modulation test can be made. In other words, the dc voltage reading corresponds to the peak carrier level, and the ac voltage reading corresponds to the rms voltage of the modulating

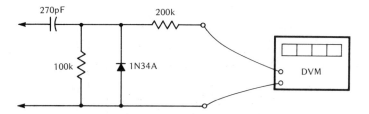

Note: *This demodulator probe does not read peak voltages directly. The DVM indicates roughly half of the peak voltage value. The value of the 200k resistor can be changed to provide a convenient scale factor.*

The probe has a carrier frequency range from 0.5 MHz to 250 MHz. It has a modulated-signal range from 30 Hz to 5 kHz.

At 0.5 MHz the probe has an input impedance of 25 kilohms, and at 200 MHz the probe has an input impedance of 2.5 kilohms.

The input peak signal level should not exceed 35 volts, to avoid possible damage to the 1N34A diode.

Figure 3-10   A demodulator probe configuration.

signal. In turn, the corresponding peak voltage of the modulating signal (assuming a sine-wave signal) is equal to 1.4 times the rms reading. The percentage modulation is equal to the ratio of peak modulating voltage to peak carrier-level voltage.

DISTORTION TEST PROBE

The distortion test probe shown in Figure 3-11 is used to make tests for sine-wave distortion, as in high-fidelity troubleshooting. It consists of a very closely matched pair of diodes and a capacitor. The sine wave to be checked for distortion is first applied between Input 1 and gnd. The VOM is operated on its dc-voltage function. This test indicates whether the positive-peak voltage is precisely the same as the negative-peak voltage in the ac waveform. If the peak voltages are equal, the VOM will indicate zero, even when operated on its lowest dc-voltage range.

Next, the sine wave to be checked for distortion is applied between Input 2 and gnd shown in Figure 3-11. The VOM is operated on its dc-voltage function. This test indicates whether the average value of the positive half cycle is precisely the same as the average value of the negative half cycle. If the average values are equal, the VOM will indicate zero, even when operated on its lowest dc-voltage range.

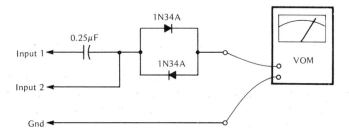

Note: *The diodes must be a closely matched pair.*

Caution: *If the waveform applied to Input 2 has a dc component, the meter indication will be false. Stated otherwise, an ac waveform has an average value of zero; it has no dc component, and it produces zero deflection when directly applied to a dc voltmeter.*

Note: *This method is qualitative, and not quantitative. To measure the percentage distortion, a harmonic distortion meter is required.*

Figure 3-11   Distortion test probe configuration.

The foregoing distortion tests are based on the fact that whenever a sine waveform is distorted, its form factor will be changed more or less. The form factor of a half cycle is equal to the ratio of its peak voltage to its average voltage. The distortion test probe depicted in Figure 3-11 employs a pair of back-to-back diodes in order to check the positive and negative peak voltages simultaneously, and to check the positive and negative average values simultaneously. A quick check provides a comparatively sensitive indication of distortion.

ONE-SHOT PEAK-HOLD UNIT

Electronic troubleshooters are occasionally concerned with surge voltages on dc supply lines, as in digital power-supply systems. Since a voltage surge may be a one-shot event, a peak-hold unit with a comparatively long time constant is required. A typical configuration, shown in Figure 3-12, consists of an 8 μF capacitor and a power-type silicon diode. The peak-hold unit is connected to the line under test with the indicated polarity. When a voltage surge occurs the DVM reading "jumps up" to the peak value of the surge. Then the DVM reading gradually decreases to the prevailing line-voltage value.

The surge peak reading gradually decreases chiefly because of the measuring current demand by the DVM. If you desire to slow down the rate of decrease, the time-constant of the measuring circuit may be increased. For example, a higher value of charging capacitance may be used, or the input resistance to the DVM may be

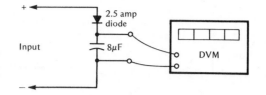

Note: *The 8 µF capacitor should have very high insulation resistance. GE Pyranol capacitors are suitable.*

The diode should have very high back resistance and a high forward current rating. The Archer 276-1114 diodes are suitable.

Note: *The diode has a barrier potential of approximately 0.25 volt. This results in a DVM reading that is about 0.25 volt less than the actual peak voltage.*

Figure 3-12   Configuration for one-shot peak-hold unit.

increased. If the capacitor value is doubled, the rate of decrease will be reduced to one-half. Similarly, if a 10-meg resistor is connected in series with the DVM, the rate of decrease will be reduced to one-half (the 10-meg resistor also causes the DVM to indicate one-half of the source voltage value).

ONE-SHOT DIP-HOLD UNIT

Electronic troubleshooters are also concerned on occasion with dip voltages on dc supply lines, as in digital power-supply systems. The voltage might drop briefly to 60 percent of normal, to 10 percent of normal, or even to zero. Since a voltage dip may be a one-shot event, a dip-hold unit with a comparatively long time constant is required.

A typical configuration, shown in Figure 3-13, consists of two 8 µF capacitors and two power-type silicon diodes. The dip-hold unit is connected to the power line under test with the indicated polarity. When a voltage dip occurs, the DVM reading "jumps up" to half the dip voltage. For example, if the normal line voltage is 12 volts and a brief dip to 10 volts occurs, the DVM will indicate 1 volt. Or, if the 12-volt line dips briefly to zero, the DVM will indicate 6 volts. Then the DVM reading gradually decays toward zero.

Operation

With reference to Figure 3-13, as long as the line voltage is steady, the input capacitor is charged to the prevailing line voltage. The

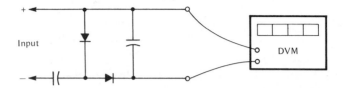

Note: Diode and capacitor characteristics are as noted in Figure 3-12.

Since each of the diodes has a barrier potential of about 0.25 volt, the DVM will indicate a voltage value that is about 0.5 volt less than the actual dip voltage.

Figure 3-13   Configuration for one-shot dip-hold unit.

output capacitor cannot charge because its associated diode has reversed polarity with respect to the line polarity. However if the line voltage suddenly dips, the input capacitor will partially discharge into the output capacitor. Because both capacitors have the same value, the input capacitor will discharge until its falling voltage equals the rising voltage of the output capacitor.

For example, suppose that the line voltage dips briefly to zero. Since the input capacitor has been standing at 12 volts, it discharges to 6 volts, and the output capacitor charges to 6 volts as the line voltage passes through zero. The 6-volt charge on the output capacitor is "trapped" and its value is indicated by the DVM. This 6-volt reading is multiplied by 2 to calculate the dip voltage of 12 volts.

As another illustration, suppose that the line voltage dips briefly to 6 volts. Since the input capacitor has been standing at 12 volts, it discharges to 9 volts, and the output capacitor charges to 3 volts as the line voltage passes through 6 volts. In other words, the 6-volt excess charge in the input capacitor will discharge until its falling voltage equals the rising voltage of the output capacitor. Or, the DVM indicates 3 volts—this 3-volt reading is multiplied by 2 to calculate the dip voltage of 6 volts.

As noted previously, the DVM reading gradually decreases because of the measuring current demand by the DVM. If you desire to slow down the rate of decrease, the capacitor values may be doubled. This modification will reduce the rate of decrease to one-half. Similarly, if a 10-meg resistor is connected in series with the DVM, the rate of decrease will be reduced to one-half (the DVM will now indicate one-fourth of the dip voltage). If both modifications are employed, the rate of decrease will be reduced to one-fourth, or, the practical hold time will be four times as long.

One-shot peak-hold and dip-hold units are also called *pulse stretchers*. In effect they function to increase the pulse width so that its peak value can be measured. The hold units described previously operate by trading off a substantial input current demand for increased output pulse width. Since power supplies, for example, have low output impedance, the input current demand of a hold unit can be supplied without objectionable loading.

As a practical observation, the dip-hold unit described above will also operate as a peak-hold unit. That is, as long as the line voltage is steady, the input capacitor in Figure 3-13 is charged to the prevailing line voltage. However if a surge pulse occurs and the line voltage increases for a brief interval, the input capacitor immediately "follows" this peak increase. Then as the surge pulse drops back to the prevailing line-voltage value, the extra charge on the input capacitor discharges into the output capacitor and half of the surge pulse voltage is indicated by the DVM.

For example, suppose that the prevailing line voltage is 12 volts and that a sudden surge pulse occurs with a peak value of 3 volts so that the line voltage rises momentarily to 15 volts. The input capacitor in Figure 3-13 then immediately charges to 15 volts. Then, as the surge pulse passes and the line voltage returns to its prevailing 12-volt level, the 3-volt excess charge on the input capacitor will discharge until its falling voltage equals the rising voltage on the output capacitor. Or, the DVM indicates 1.5 volts—this 1.5-volt reading is multiplied by 2 to calculate the peak surge voltage of 3 volts.

Of course, a dip-hold unit does not indicate whether a surge or a dip has been "caught." Therefore, if the troubleshooter does not know whether the line is subject to surges, or to dips, or to both, he should monitor the line with both a peak-hold unit and a dip-hold unit. Since a peak-hold unit does not respond to a brief dip in line voltage, the troubleshooter can observe whether a DVM reading at the dip-hold unit actually indicates a dip voltage, or whether it may be indicating a surge voltage. The troubleshooter then can easily make a complete checkout of a line that is subject to both voltage surges and dips.

CHAPTER 4

DC CURRENT TESTS AND MEASUREMENTS

CONSTANT-CURRENT VS. OHMIC-CURRENT SOURCES • TRANSISTOR BETA MEASUREMENT • AMPLIFIER CURRENT GAIN • CURRENT PEAK-HOLD UNIT • CURRENT DIP-HOLD UNIT • HOW TO ELIMINATE THE VOLTAGE BURDEN OF A CURRENT SHUNT RESISTOR • PULSATING DC CURRENT MEASUREMENTS • EXPERIMENTAL PROJECTS

CONSTANT CURRENT VS. OHMIC CURRENT SOURCES

Electronic troubleshooters are often concerned with constant-voltage sources. For example, a regulated power supply is a practical constant-voltage source—the source voltage remains constant when the load resistance is varied. Electronic troubleshooters are also concerned with constant-current sources. For example, a silicon transistor with a fixed base-emitter bias is a practical constant-current source—the source current remains constant as the load resistance is varied.

Again, the troubleshooter is often concerned with sources that are neither constant-voltage sources nor constant-current sources—for example, an ordinary power supply with an RC filter has an output voltage that varies when the load resistance is varied. The ordinary power supply also has an output current that varies when the load resistance is varied. Therefore, to avoid confusion in the following discussion, sources such as ordinary power supplies will be termed *ohmic current sources.*

In the present state of the art, electronic troubleshooters are becoming more involved with constant-current sources. For example, a silicon transistor with a fixed base-emitter bias is a practical constant-current source (see Figure 4-1). Otherwise stated, if the constant-current source works into a 10-kilohm load, it will supply the same

59

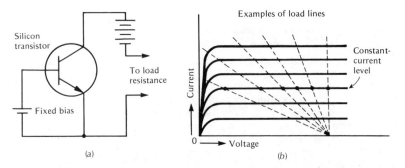

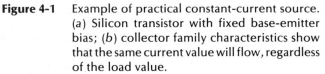

(a)

(b)

Figure 4-1   Example of practical constant-current source. (a) Silicon transistor with fixed base-emitter bias; (b) collector family characteristics show that the same current value will flow, regardless of the load value.

amount of current as if it works into a 10-ohm load, or into a 0-ohm load. (This circuit is extensively used in semiconductor configurations.)

It is apparent in Figure 4-1(b) that the same value of current will flow, regardless of the load resistance that is used. However the voltage drop across the load is proportional to its resistance value. In other words, a *constant-current* source is an *ohmic voltage* source. The basic constant-current arrangement shown in Figure 4-1 is often encountered in electronic circuitry.

For purposes of practical constant-current tests and measurements the electronic troubleshooter uses a comparatively simple type of constant-current source, as shown in Figure 4-2. This is a practical source of constant current because the battery voltage is chosen to be much higher than any of the voltages in the circuit under test. Accordingly, the internal resistance of the constant-current source is much higher than any of the resistances in the circuit under test, and a constant-current value will be injected at any practical test point in the circuit.

A constant-current technique for measurement of transresistance was noted in Chapter 1. Various other constant-current tests can be used to speed up troubleshooting procedures.

TRANSISTOR BETA MEASUREMENT

A constant-current test for measurement of transistor beta is helpful in practical work, as shown in Figure 4-3. With the switch in

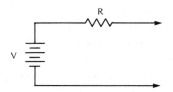

Note: When the values of R and V are comparatively high, this arrangement serves as a practical constant-current source.

For example, if V = 45 volts, and R = 4.5 megohms, a constant current of 10 microamperes is supplied.

Figure 4-2   A practical constant-current source.

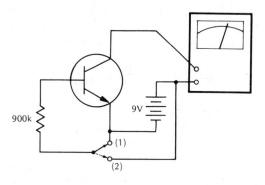

Note: When switch is thrown from position (1) to position (2), the beta value is equal to 0.1 of the microammeter scale reading.

Figure 4-3   Constant-current measurement of transistor beta value.

position (1), the leakage current of the transistor is normally unreadable on the microammeter scale of the current meter. When the switch is thrown to position (2), the beta value of the transistor is equal to 0.1 of the microammeter scale reading.

Since a 900-kilohm resistor is connected in series with the 9-volt battery, and the transistor base input resistance is comparatively low, the troubleshooter knows that 10 μA of base current are being injected. Accordingly, the collector-current flow is equal to beta times 10 μA. Note that if a readable value of collector-current flow occurs with the switch in its (1) position, the transistor is defective and should be discarded. A measured beta value lower than the rated value also indicates that the transistor is defective.

Example: A silicon transistor has no readable leakage current when the 900-kilohm base resistor is returned to the emitter terminal. When the base resistor is returned to the 9-volt battery terminal, the microammeter reads 1250 microamperes, and the transistor beta value is 125—a reasonable normal value.

AMPLIFIER CURRENT GAIN

A constant-current test for measurement of amplifier current gain is also helpful in practical troubleshooting procedures. The procedure is as follows:

1. With reference to Figure 4-4, a constant current of 10 μA is injected into the base circuit of the amplifier transistor.
2. A dc voltmeter is connected across the collector load resistor to measure the voltage change across the resistor when the constant-current voltage is applied.
3. In this example, the voltage drop across the load resistor changed from 4 volts to 5 volts when 10 μA were injected into the base circuit.
4. In accordance with Ohm's law, a 1-volt change across 65 kilohms corresponds to a current change of 154 μA.
5. The current gain of the amplifier was therefore 15.4 times in this example.

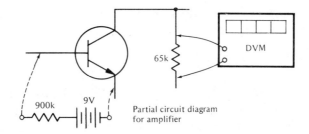

Use CAT! Punch out Ohm's-law answers on your pocket calculator. Calculator-aided troubleshooting can save valuable time on many Ohm's-law calculations.

Figure 4-4   Measurement of amplifier current gain.

CURRENT PEAK-HOLD UNIT

It is occasionally necessary to measure the peak current demand of a utilization device or system. For example, intermittent fuse-blowing can be a "tough dog" problem, and realistic analysis requires measurement of the peak current demand. With reference to Figure 4-5, a current peak-hold unit consists of a suitable shunt with a voltage peak-hold arrangement.

Note in Figure 4-5 that the voltage drop across the shunt resistor diminishes the voltage that is otherwise available to the load. This voltage drop across the shunt is called the *voltage burden* of the current peak-hold unit. Good practice dictates that the shunt value impose as small a voltage burden as is feasible. On the other hand, if an excessively small value of shunt resistance is used, it will be difficult or impossible to accurately measure the prevailing current and surge current values.

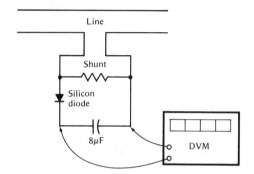

Note: *A 1-ohm shunt provides a readout of 1 ampere per volt. A 1-kilohm shunt provides a readout of 1 milliampere per volt.*

Shunts must have adequate power ratings. For example, a current flow of 5 amperes in 1 ohm dissipates 25 watts.

Note: *The silicon diode has a barrier potential of approximately 0.6 volt. From a practical viewpoint, this means that the DVM will indicate about 0.6 volt less than the actual voltage drop across the shunt.*

Figure 4-5   Current peak-hold arrangement.

CURRENT DIP-HOLD UNIT

Power-supply troubleshooting, as in digital systems, is occasionally concerned with current dips in the supply line. The troubleshooter needs to use a current dip-hold unit in order to measure how much the current dips momentarily. With reference to Figure 4-6, a current dip-hold unit consists of a suitable shunt with a voltage dip-hold arrangement. It should be noted that the dip-hold arrangement cannot distinguish between surges and dips. Therefore it is often desirable to employ both a peak-hold unit and a dip-hold unit. Note that the same shunt resistor can be used to energize both

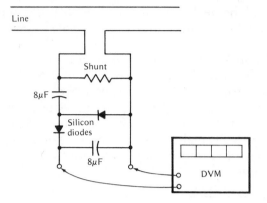

Note: *Conventional silicon diodes have a barrier potential of approximately 0.6 volt. As a result, the DVM indicates about 1.2 volts less than the actual dip voltage.*

Figure 4-6 Current dip-hold arrangement.

the peak-hold unit and the dip-hold unit, thereby minimizing the voltage burden.

HOW TO ELIMINATE THE VOLTAGE BURDEN OF A CURRENT SHUNT RESISTOR

It was previously explained how the troubleshooter can use a bias box to eliminate the current demand of a dc voltmeter. Next, if we apply the principle of duality, we will recognize that a related procedure can be employed to eliminate the voltage burden of a current shunt resistor. This procedure permits the troubleshooter to make accurate dc current measurements in circuits that would otherwise be objectionably disturbed by the voltage burden of a dc current meter, or by the voltage burden of a corresponding current shunt.

With reference to Figure 4-7, the voltage burden eliminator consists of a DVM, battery, variable resistor, and a current meter. The DVM functions as a sensitive null indicator in this application. Note that the measuring circuit applies a cancellation voltage across the shunt resistor. When the DVM indicates zero, the zero-adjust control is set to precisely cancel out the voltage drop that would otherwise occur across the shunt resistor; the current meter then accurately indicates the line current flow.

If a large value of line current is to be measured, the battery in Figure 4-7 may be replaced by a heavy-current power supply. Both the

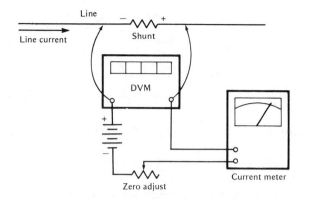

Line

Line current

Shunt

DVM

Zero adjust

Current meter

Figure 4-7   Voltage burden eliminator for current shunt resistor.

shunt resistor and the zero-adjust resistor must have adequate power ratings. Observe that when the voltage drop across the shunt resistor is precisely nulled, the line current is effectively bypassed around the shunt resistor via the measuring circuit, or, the power supplied by the battery effectively reduces the value of the shunt resistance to zero.

PULSATING DC CURRENT MEASUREMENTS

Many types of pulsating dc current are encountered in electronic troubleshooting procedures. *Pulsating* means that the value of the dc current is changing, although the current never changes polarity. One important example is the rectified dc current output from a diode in a power supply. This pulsating dc current has the form of half sine waves, as depicted in Figure 4-8. When this rectified sine-wave current from the diode flows through a dc current meter, the scale indication is 0.318 of peak value (average value).

In some applications the peak value is of concern; this peak current value is equal to 3.14 times the current meter reading. Again, the rms value of the rectified sine wave is of concern in other applications. This rms value is equal to one-half of peak current, or the rms value is equal to 1.57 times the current meter reading. Note that the rms value of the rectified sine wave has the same power capability as a steady dc current of the same value.

As an example of power capability, if a diode lamp dimmer supplies a dc current of 0.5 ampere as measured on a dc current meter, the rms value of this current is 0.785 ampere. In other words, when this 0.785 ampere rms flows through a lamp filament, the temperature of

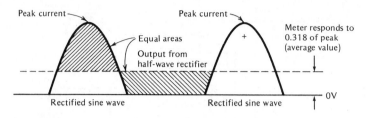

Figure 4-8   Rectified sine-wave parameters.

the filament will be the same as if a pure dc current of 0.785 ampere were flowing through the filament.

Note that the numerical relations noted above apply only to pulsating dc current that has a half-sine waveform. Other numerical relations apply to pulsating dc current waveforms that have pulse waveshapes, for example. The same general relations that apply to dc voltage pulses apply to dc current pulses. Thus, the pulse width in a dc current pulse train can be measured with a current meter in the same general way that the pulse width in a dc voltage pulse train is measured, as is explained in Chapter 2.

EXPERIMENTAL PROJECTS

Construction and Checkout of a One-Transistor Class-A Amplifier

The configuration of a one-transistor Class-A amplifier used in this project is shown in Figure 4-9. The amplifier is constructed from the following parts:

- Wood or plastic base 4″ × 1¼″ × ⅛″
- (2) 8-lug terminal strips
- (2) ⅜″ 6-32 flat-head machine screws and nuts
- (2) 100 μF 35V electrolytic capacitors
- 20K ⅛W resistor
- 100K ⅛W resistor
- 150K ⅛W resistor
- 65K ⅛W resistor
- NPN silicon transistor
- 9V battery
- Battery connector
- Insulated hookup wire (6 in.)

The two terminal strips are mounted symmetrically on the base with their angle brackets overlapping, and secured by the machine

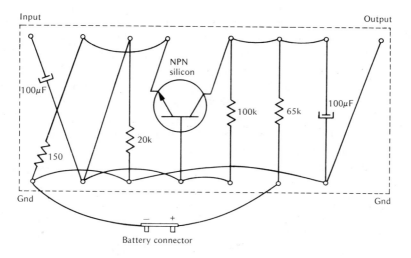

Input   Output

Figure 4-9   Breadboarded one-transistor Class-A amplifier.

screws. The "breadboard" consists of two rows of terminal lugs with
½-inch spacing between them. Parts are mounted between the
terminal lugs as shown in Figure 4-9, and are supported by their leads.
Interconnections are made with the hookup wire.

When construction is completed, *measure the dc voltages* in the
amplifier circuit. These measured values will depend to some extent
on the tolerances of the resistors, the characteristics of the particular
transistor, the battery terminal voltage under load, and the type of
voltmeter that is used. Typical dc voltage values are indicated in
Figure 4-10.

Next *measure the amplifier input resistance.* Proceed as follows:

1. Measure the dc voltage at the base of the transistor.
2. Shunt a 50 kilohm resistor from the base terminal to ground
   and note the resulting reduction in base voltage.
3. Calculate the current flow through the 50 kilohm resistor, and
   divide this current value into the base-voltage change.
4. The quotient is equal to the amplifier input resistance; a typical
   value for the configuration in Figure 4-11 is 1800 ohms.

Now proceed to measure the *amplifier output resistance,* as
follows:

1. Perform the test described in Chapter 1 under "Dynamic
   Internal Resistance." This test is made at the collector terminal
   of the transistor; therefore the dynamic internal resistance

will be the same as the output resistance of the amplifier in this configuration.

2. We recognize that the foregoing equivalence holds true in this example because the amplifier under test is unloaded (is not driving a following stage, or a speaker). On the other hand, if the amplifier were loaded, its output resistance could not be determined by measuring the dynamic internal resistance from the collector terminal. Stated otherwise, although the amplifier output resistance remains unchanged whether the amplifier is unloaded or loaded, the amplifier dynamic internal resistance measured from the collector terminal is changed as a result of loading.

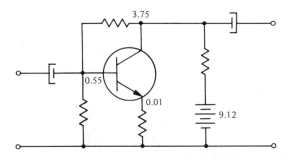

Figure 4-10   Typical dc voltages in experimental amplifier circuit, measured with a DVM.

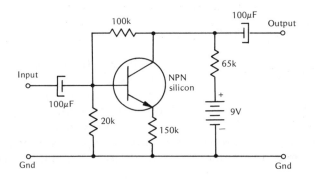

Note: *The voltage gain of the amplifier depends upon the value of the feedback resistor. An adjustable-gain amplifier can be obtained by using a 200k potentiometer in place of the 100k fixed resistor.*

Figure 4-11   Configuration for a one-transistor Class-A amplifier.

Next, it is interesting to measure the *transresistance* of the configuration in Figure 4-11. This particular transresistance test is a *constant-current* test, and is made as shown in Figure 4-12. Transresistance is measured in ohms; it is the ratio of an input current change to the corresponding output voltage change. The procedure is as follows:

1. Construct a 10-μA constant-current source by connecting a 900,000-ohm resistor in series with a 9V battery.
2. Connect a DVM from the collector to ground in the amplifier, and measure the dc voltage.
3. Connect the constant-current source from base to ground in the amplifier, and note the change in collector dc-voltage reading.
4. Divide the change in collector dc-voltage reading by 10 μA to calculate the transresistance of the amplifier. For example, if the collector dc-voltage value changes from 3.67 volts to 2.75 volts when the constant-current source is applied, the transresistance is equal to $^{0.92}\!/_{10} \times 10^6$, or 92,000 ohms. Note that if the constant-current direction is reversed, the collector dc voltage will increase when the constant-current source is applied. However, if the amplifier is operating in Class A, the same value of transresistance will be measured.

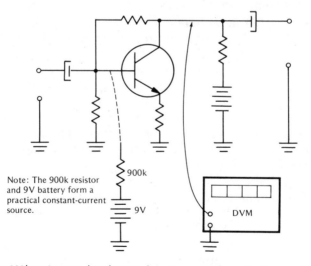

Note: The 900k resistor and 9V battery form a practical constant-current source.

900k

9V

DVM

Note: *The 900k resistor and 9V battery form a practical constant-current source.*

Figure 4-12   Check of amplifier transresistance.

Now observe the following practical examples of *circuit loading*. With reference to Figure 4-11, proceed as follows:

1. Measure the base voltage with a 1000 ohms/volt meter, with a 20,000 ohms/volt meter and with a DVM.
2. Compare the three dc voltage values that you measured.
3. Measure the collector voltage with a 1000 ohms/volt meter, with a 20,000 ohms/volt meter, and with a DVM.
4. Compare the three dc voltages that you measured.
5. Measure the emitter voltage with a 1000 ohms/volt meter, with a 20,000 ohms/volt meter and with a DVM.
6. Compare the three dc voltage values that you measured.

Next, observe the following practical examples of *in-circuit resistance measurement*. With reference to Figure 4-11, proceed as follows:

1. Measure the resistance from base to ground with a hi-pwr ohmmeter; connect the positive lead of the ohmmeter to the base terminal. (The 9V battery *must* be disconnected.)
2. Repeat the resistance measurement with the negative lead of the ohmmeter connected to the base terminal.
3. Measure the resistance from base to ground with a lo-pwr ohmmeter.
4. Compare the three resistance values that you measured.
5. Measure the resistance from collector to ground with a hi-pwr ohmmeter; connect the positive terminal of the ohmmeter to the collector terminal.
6. Repeat the resistance measurement with the negative lead of the ohmmeter connected to the collector terminal.
7. Compare the three resistance values that you measured.
8. Measure the resistance from emitter to ground with a hi-pwr ohmmeter; connect the positive terminal of the ohmmeter to the emitter terminal.
9. Repeat the resistance measurement with the negative lead of the ohmmeter connected to the emitter terminal.
10. Compare the three resistance values that you measured.

Now measure the *amplifier current gain,* as shown in Figure 4-4. A typical current gain for this configuration is 15 times. Then measure the *amplifier voltage gain.* The voltage gain is measured as follows:

1. With reference to Figure 4-11, measure the base voltage to ground and measure the collector voltage to ground.
2. Shunt a 0.25 megohm potentiometer across the 20-kilohm base resistor, and adjust the potentiometer for a reduction of 10 mV in the base voltage.
3. Note the corresponding increase in collector voltage that results.
4. The voltage gain of the amplifier is equal to the output voltage change divided by the input voltage change. A typical voltage-gain value for this configuration is 45 times.

The *amplifier power gain* can now be calculated—the power gain is equal to the product of the current gain and the voltage gain. A typical power gain for this configuration is 675 times. This is a power gain of 29 dB.

CHAPTER 5

AC VOLTAGE TESTS AND MEASUREMENTS

SINE-WAVE VS. COMPLEX-WAVE MEASUREMENTS • DISTORTION CHECK-ING WITH AN AC VOLTMETER • TURNOVER CHECK • PEAK VOLTAGE MEASUREMENTS OF AC PULSES VS. DC PULSES • TRUE RMS VALUES • SAW-TOOTH AVERAGE VALUE AND APPARENT RMS VALUE • APPARENT RMS VALUES OF BASIC PULSATING DC AND AC WAVEFORMS • PEAK-RESPONSE AC VOLTMETER • AC WAVEFORMS WITH DC COMPONENTS • SELECTION OF A BLOCKING CAPACITOR • QUICK TEST FOR CIRCUIT LOADING AND/OR INSUFFICIENT BLOCKING CAPACITANCE • AC VOLTMETER FRE-QUENCY CAPABILITIES • AC VOLTAGE MEASUREMENTS IN LOW-LEVEL CIRCUITS

SINE-WAVE VS. COMPLEX-WAVE MEASUREMENTS

Electronic troubleshooters know that there are basic distinctions between sine-wave and complex-wave parameters. These distinctions must be taken into account for evaluation of ac voltmeter readings. For example, the sine wave and the square wave depicted in Figure 5-1 have the same peak-to-peak voltage. However, the sine wave has an rms voltage equal to 0.707 of peak, whereas the square wave has an rms voltage equal to peak voltage. Although the rms value of the sine wave can be directly measured with a service-type ac voltmeter, the rms value of the square wave cannot be directly measured with a service-type ac voltmeter.

Nearly all service-type ac voltmeters employ half-wave instrument rectifiers. Accordingly, the ac voltmeter responds to the average value of one-half cycle in the applied ac waveform. With reference to Figure 5-2, the average value of a sine-wave half cycle is 0.318 of peak voltage, whereas the average value of a square-wave half cycle is 0.5 of peak voltage.

The scale of a service-type ac voltmeter is usually calibrated to indicate the rms values of sine waveforms. The significance of the

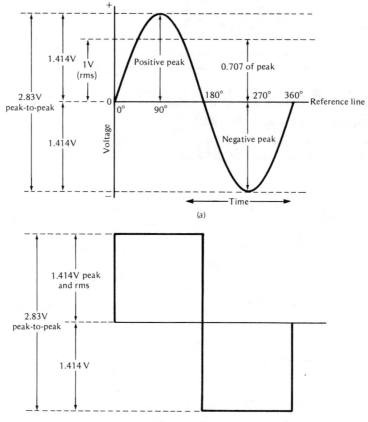

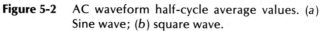

Figure 5-1 AC waveform parameters. (a) Sine wave; (b) square wave.

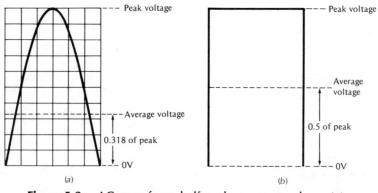

Figure 5-2 AC waveform half-cycle average values. (a) Sine wave; (b) square wave.

foregoing parameters is that *a service-type ac voltmeter will not indicate the rms value of a square wave.* In other words, the rms value of a sine wave is equal to 0.707 of peak. Since the ac voltmeter responds to the half-cycle average value, or 0.318 of peak, the scale is calibrated to indicate 2.22 times the average value of the sine wave. Now, if the troubleshooter applies a square-wave input to the ac voltmeter, *the scale indication will be 1.11 times the peak voltage of the square wave.*

In the case of a square wave, *the rms voltage is equal to the half-cycle peak voltage,* as shown in Figure 5-2. Therefore, to calculate the rms voltage of the square wave, *the troubleshooter must take 0.9 of the scale reading.*

DISTORTION CHECKING WITH AN AC VOLTMETER

Suppose that an amplifier develops clipping distortion, as shown in Figure 5-3. The ac voltmeter then "sees" a half-sine wave with more or less of its peak excursion clipped off. This waveform is not a pure half-sine wave, and although the waveform has flat top and bottom intervals it is not a pure square wave (because its sides slope rather than being vertical).

The practical result is that the ac voltmeter indicates a higher voltage than for a pure sine wave—although the indicated voltage is less than for a pure square wave. The troubleshooter can check for sine-wave clipping as follows:

1. Measure the peak voltage of the waveform with a peak-reading probe or with a peak-hold unit.
2. Measure the apparent rms voltage of the waveform on the ac-voltage function of the meter.
3. Then, if the rms reading is equal to 0.707 of the peak value, the troubleshooter concludes that the waveform is a pure sine wave.
4. On the other hand, if the rms reading is greater than 0.707 of the peak value, the troubleshooter concludes that the original sine waveform has been clipped.

TURNOVER CHECK

Turnover is often encountered in troubleshooting procedures. *Turnover occurs in a test if the meter reading changes when the test leads are reversed.* As an illustration, consider peak-voltage measure-

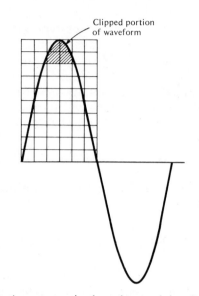

Note: *When peak-clipping occurs, the form-factor of the clipped half-sine wave is changed, and this change, if substantial, can be determined by voltmeter tests. This is a qualitative test, not quantitative. To measure percentage harmonic distortion, a harmonic distortion meter must be used.*

Figure 5-3   Example of clipping distortion in a sine wave.

ments of the pulse waveform in Figure 5-4. When a peak-reading probe is applied, we might measure the positive-peak voltage. Then, when the probe leads are reversed, we will measure the negative-peak voltage. Since the positive-peak voltage is greater than the negative-peak voltage, we have observed turnover action.

Next, if we measure the apparent rms voltages for the positive and negative portions of the pulse waveform in Figure 5-4, we will find that these two values are the same—there is no turnover. To understand why no turnover is encountered when we measure the apparent rms voltages of the waveform on the ac function of a multimeter, note the following facts:

1. An ac voltmeter with a half-wave instrument rectifier responds to the average value of the positive excursion, or to the average value of the negative excursion, in the waveform under test.
2. Since the average value of the positive excursion of this waveform is the same as the average value of the negative excursion, the ac voltmeter reading is the same when its test leads are reversed.

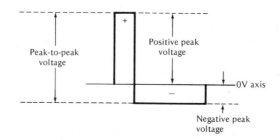

Note: The average value of the positive excursion is the same as the average value of the negative excursion.

Figure 5-4   Basic ac pulse waveform.

Apparent RMS Readings

The scale of an ac voltmeter is calibrated in rms volts for a pure sine wave. When we measure the voltage of the positive excursion in Figure 5-4, we obtain the *apparent rms* reading. This apparent rms reading is equal to 2.22 times the average value of the positive excursion (since service-type ac voltmeters employ half-wave instrument rectifiers).

Similarly, when we measure the voltage of the negative excursion in Figure 5-4, we obtain the apparent rms reading for the negative portion of the waveform. *This apparent rms reading is the same as measured for the positive excursion, because the average values of the excursions are the same.*

PEAK VOLTAGE MEASUREMENTS OF AC PULSES VS. DC PULSES

Important point: Troubleshooters encounter both ac pulses and dc pulses in electronic circuitry, and peak-voltage measurements are different for these related types of waveforms. The essential points to keep in mind are:

1. A peak-reading probe measures the peak voltage of an ac pulse waveform; turnover occurs when the probe leads are reversed.
2. A peak-reading probe measures the peak voltage of the *ac component* in a corresponding dc pulse. The complete peak voltage is equal to the *sum* of the probe output voltage and the average value of the dc pulse, as measured with a dc voltmeter.

3. To distinguish between an ac pulse train and a dc pulse train when troubleshooting pulse circuitry, note that a peak-reading probe will give a zero reading in a turnover test of a dc pulse train.

TRUE RMS VALUES

Electronic troubleshooters are infrequently concerned with the true (actual) rms values of complex waveforms. Note that true rms ac voltmeters may be used when true rms values of complex waveforms are of concern. However, a true rms meter is considerably more costly than a meter that employs an instrument rectifier.

SAWTOOTH AVERAGE VALUE
AND APPARENT RMS VALUE

Sawtooth waveforms are often encountered in electronic circuitry. In Figure 5-5, the half-cycle average value of the sawtooth waveform is 0.25 of its peak value. This average value is evident, inasmuch as the average value of a half cycle in a square wave is 0.5, and the sawtooth half cycle has one-half the area of a square-wave half cycle. *The average value of a half cycle in a complex ac waveform is proportional to its area, regardless of its waveshape.* This law follows from the fact that the amount of electricity in a half cycle is proportional to its area.

Inasmuch as the average value of a half cycle in a sawtooth waveform is 0.25 of peak, its *apparent rms voltage* is equal to 0.555 of

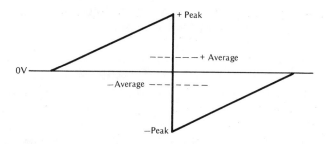

Note: *This is a symmetrical waveform because the positive and negative excursions have the same shape.*

Its + average value is 0.25 of peak, and its − average value is also 0.25 of peak.

Figure 5-5　Peak and average values in an ac sawtooth waveform.

peak, as measured on an ac voltmeter that employs a half-wave instrument rectifier. Observe that if this ac sawtooth waveform were applied to a dc voltmeter, the reading would be zero. On the other hand, if a diode were connected in series with the dc voltmeter lead, the reading would be 0.25 of peak.

Important point, repeated: Although an ac voltmeter employs a half-wave instrument rectifier, and a half-wave rectifier was used with a dc voltmeter in the foregoing example, the scale readings were greatly different. The ac voltmeter reading is greater than the dc voltmeter reading because the scale on the ac voltmeter is calibrated to read 2.22 times the average value of the half-rectified waveform.

Triangular Waveform

Triangular waveforms (Figure 5-6) are occasionally encountered in electronic circuitry. A triangular waveform is basically two sawtooth waveforms back to back. The half-cycle area of a triangular wave is twice the area of the reference sawtooth wave. This does not mean, however, that the average value of the triangular wave is double the average value of the reference sawtooth wave—*when two sawtooth waveforms are placed back-to-back, the waveform period is thereby doubled.* Accordingly, the average half-cycle value of a triangular waveform is 0.25, as for a sawtooth waveform.

Reminder: The average half-cycle value of any waveform is equal to the ratio of its area to its period. If you visualize a "filled-in" area over

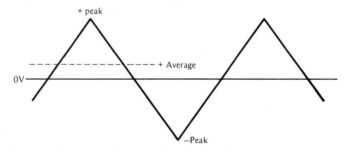

Note: *A triangular waveform has a half-cycle average value equal to 0.25 of peak. This is the same average value as for a sawtooth waveform. We will observe that the half-cycle area of a triangular wave is double the half-cycle area of a sawtooth wave—but the period of a triangular wave is double that of the reference sawtooth wave.*

Figure 5-6   Peak and average values in an ac triangular waveform.

the entire period of a triangular wave, you will see that the area of its half-cycle excursion is 0.25 of this maximum possible ("filled-in") area.

APPARENT RMS VALUES OF BASIC
PULSATING DC AND AC WAVEFORMS

Electronic troubleshooters are concerned with both pulsating dc waveforms and with ac waveforms. A pulsating dc waveform has an apparent rms value, just as an ac waveform has an apparent rms value. In Figure 5-7, examples of both waveform categories are shown. Thus, the square wave is an ac waveform, the sawtooth wave is an ac waveform, the half-rectified sine wave is a pulsating dc waveform, and the full-rectified sine wave is a pulsating dc waveform.

As explained above, the apparent rms value of the ac square wave is 1.11 times its peak value, and the apparent rms value of the ac sawtooth wave is 0.555 times its peak value. Next we observe that the apparent rms value of a half-rectified sine waveform is 0.707 times its peak value, and the apparent rms value of a full-rectified sine wave is 1.414 times its peak value, as indicated by an ac voltmeter that employs a half-wave instrument rectifier. Note that these are the logical apparent rms indications, based on the instrument circuitry.

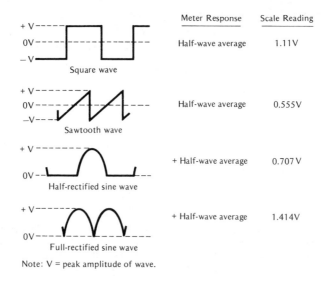

	Meter Response	Scale Reading
Square wave	Half-wave average	1.11V
Sawtooth wave	Half-wave average	0.555V
Half-rectified sine wave	+ Half-wave average	0.707 V
Full-rectified sine wave	+ Half-wave average	1.414V

Note: V = peak amplitude of wave.

Figure 5-7 Apparent rms values of basic ac and pulsating dc waveforms.

Of course, a turnover check with the ac voltmeter leads will give the same apparent rms reading for the ac waveforms, but a turnover check will give a zero reading for the pulsating dc waveforms. Note that the half-rectified sine waveform and the full-rectified sine waveform have their related ac waveforms—if the pulsating dc waveform is coupled to the ac voltmeter via a blocking capacitor, lower apparent rms values will be indicated. *It is important to keep this fact in mind if you are using an ac voltmeter on its output function.*

Full-Wave Instrument Rectifiers

Occasionally the troubleshooter may use an ac voltmeter that employs full-wave instrument rectifiers. In such a case, the foregoing apparent rms readings will be the same for the complex ac waveforms. On the other hand, the foregoing apparent rms readings will be different for the pulsating dc waveforms. Thus, with full-wave instrument rectifiers, an ac voltmeter will indicate an apparent rms value of 0.354 for the half-rectified sine wave, and will indicate an apparent rms value of 0.707 for the full-rectified sine wave.

These are the logical apparent rms readings, based on the instrument circuitry. Note also that a turnover test is meaningless when the ac voltmeter employs full-wave instrument rectifiers because the same reading will always be obtained, regardless of the waveform.

PEAK-RESPONSE AC VOLTMETER

A few ac voltmeters utilize peak response, with the scale calibrated in rms values. In such a case the troubleshooter must remember that apparent rms values are generally different. Thus, a peak-response ac voltmeter indicates 0.707 of peak for an ac square wave; it indicates 0.707 of peak for an ac sawtooth wave; it indicates 0.707 of peak for a rectified sine wave; and it indicates 0.707 of peak for a full-rectified sine wave.

As would be expected, a peak-response ac voltmeter will indicate the same value on a turnover test for ac waveforms, and it will indicate zero on a turnover test for a pulsating dc waveform.

True RMS AC Voltmeter

If the troubleshooter is using a true rms ac voltmeter, basic ac and pulsating dc waveforms will measure their true (actual) rms values, as exemplified in Figure 5-8. In general, the true rms value of a complex ac waveform or of a pulsating dc waveform is considerably different from its apparent rms value as indicated on a service-type ac voltmeter. This difference of indication is due to the fact that ac voltmeters with instrument rectifiers respond to the average value of a half cycle, which is a linear function of area; on the other hand, true rms voltmeters respond to the square root of the sum of the squares in the waveform (a square function).

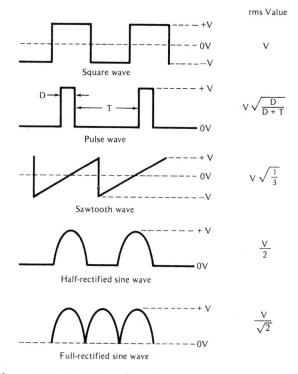

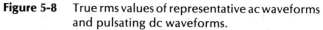

Figure 5-8   True rms values of representative ac waveforms and pulsating dc waveforms.

AC WAVEFORMS WITH DC COMPONENTS

Electronic troubleshooters often encounter ac waveforms with dc components. In Figure 5-9, an ac waveform has an average value of zero, as measured on a dc voltmeter. On the other hand, a pulsating dc waveform has a dc component that is at least equal to (but which is often greater than) the peak value of its ac component. Finally, an ac

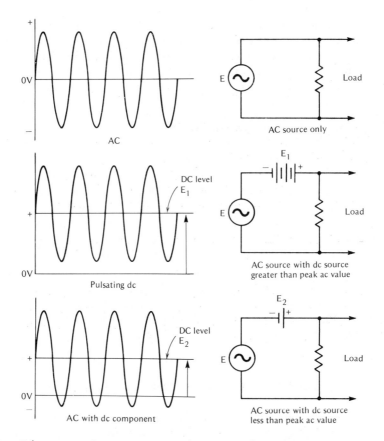

Note: When measuring component voltages in pulsating dc waveforms, or ac waveforms with dc components, the peak value of the waveform should not exceed the range limit of the voltmeter. Select a voltage range such that the peak value of the waveform is within the range. The frequency of the ac component should not exceed the frequency capability of the voltmeter.

Figure 5-9 Basic examples of ac waveform, pulsating dc waveform, and ac waveform with dc component.

waveform with a dc component has a dc component value that is less than the peak value of its ac component.

For example, the secondary of a power transformer supplies a sine-wave ac voltage. The collector of a transistor in a conventional amplifier supplies a pulsating dc voltage. The collector of a transistor in an overdriven conventional amplifier supplies an ac-with-dc-component voltage.

Important point, repeated: A peak-reading probe will indicate the $+$ peak value (or the $-$ peak value) of the ac component—the probe rejects any dc voltage that may be present. A dc voltmeter will indicate the value of any dc voltage that may be present—the dc voltmeter rejects any ac component.

AC Voltmeter Response to Pure DC Voltage

An ac voltmeter that employs instrument rectifiers responds to an applied pure dc voltage in accordance with its instrument circuitry. Thus, if an ac voltmeter with a half-wave instrument rectifier is connected to a 1.5V dc source, the scale indication will be 3.33V rms. Or, if an ac voltmeter with a full-wave instrument rectifier is connected to a 1.5V dc source, the scale indication will be 1.665V rms.

The ac voltmeter with a half-wave instrument rectifier will indicate zero on turnover. However the ac voltmeter with a full-wave instrument rectifier will indicate the same value on turnover. If either voltmeter is operated on its output function, the scale reading will be zero. Similarly, if a blocking capacitor is connected in series with either voltmeter, the scale reading will be zero.

SELECTION OF A BLOCKING CAPACITOR

When selecting a blocking capacitor for use with an ac voltmeter, the troubleshooter should use a capacitor with adequate working-voltage rating, taking into account the peak voltage that must be withstood. The capacitance of the blocking capacitor will depend upon the type of ac voltmeter that is used, the lowest frequency of test, and the indication accuracy that is desired.

The reactance of the blocking capacitor at the lowest frequency of test may be 5 percent of the meter input resistance, for example. At 1 kHz, a 0.1 μF blocking capacitor has a reactance of approximately 1500 ohms; at 100 Hz, the blocking capacitor will have a reactance of approximately 15 kilohms; at 10 kHz, the capacitor will have a reactance of approximately 750 ohms.

If a 20,000 ohms/volt ac voltmeter is used, its input resistance on its 10V range will be 200 kilohms; its input resistance on its 50V range will be 1 megohm. Or, if a DVM is used, its input resistance on all ac ranges is typically rated at 10 megohms. However, if a 1000 ohms/volt voltmeter is used, its input resistance on its 10V range will be 10 kilohms; its input resistance of its 50V range will be 50 kilohms.

QUICK TEST FOR CIRCUIT LOADING AND/OR
INSUFFICIENT BLOCKING CAPACITANCE

If the circuit under test is not seriously loaded, and if the blocking capacitor has adequate capacitance, an ac voltmeter will read practically the same voltage value when switched to an adjacent range.

Example: A 20,000 ohms/volt ac meter was operated on its 10V range with a 0.01 μF blocking capacitor. When an ac signal voltage was applied, the scale reading was 5V rms. However, when the meter was switched to its 50V range, the scale reading was 7V rms.

This difference in readings could have been caused by circuit loading, by inadequate blocking capacitance, or by a combination of both factors. To determine the cause of the discrepancy, a 0.05 μF blocking capacitor was substituted for the 0.01 μF capacitor. Then the scale reading was practically 7V rms on either the 10V range or on the 50V range. The conclusion was that circuit loading was insignificant, but that the 0.01 μF blocking capacitor had insufficient capacitance at the frequency of test.

AC VOLTMETER FREQUENCY CAPABILITIES

Many service-type ac voltmeters provide satisfactory indication accuracy over the audio frequency range from 20 Hz to 20 kHz. However some meters are rated for indication accuracy up to only 500 Hz. A few service-type meters are rated for reasonable indicating accuracy up to 100 kHz. Therefore the electronic troubleshooter must be alert to this possible limitation on ac voltmeter application.

From a practical point of view, electronic troubleshooters should use peak-reading probes for ac voltage measurements, particularly at radio and television signal frequencies. The probes that were described in Chapter 3 can be used with confidence at frequencies up to 10 MHz.

For use at frequencies from 10 MHz to 200 MHz, an isolating resistor should be connected at the peak-probe output, as depicted in Figure 5-10. The isolating resistor functions to minimize the ac component that gains entry to the test leads. As a result, measuring errors due to standing waves are avoided and the high-frequency capability of the probe is greatly extended.

Remember that the isolating resistor reduces the indicated peak voltage by a factor that depends upon its resistance value with respect to the meter input resistance. Therefore, keep this scale correction

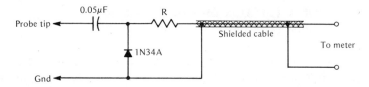

Note: A 1N34A diode has zero response to ac voltages with a peak value less than 200 mV. However this limitation can be largely overcome by applying a small forward bias to the diode, as explained in Chapter 3.

Note: When used with a TVM or DVM with 10 megohms input impedance, the value of isolating resistor R may be 100 kilohms. In turn, the peak voltage indicated on the dc function of the meter must be multiplied by a scale factor of 1.01 to obtain a precise value. (In many cases, this 1 percent error is ignored.)

A shielded input cable is helpful to avoid indication error due to stray-field pickup.

Figure 5-10   Peak-reading probe with isolating resistor for high-frequency ac voltage measurements.

factor in mind when using a peak-reading probe with an isolating resistor.

AC VOLTAGE MEASUREMENTS IN LOW-LEVEL CIRCUITS

The electronic troubleshooter needs to make ac voltage measurements in low-level circuits on occasion. Although a DVM can measure ac voltages in the lower mV range, accuracy becomes questionable, and it is impossible to measure ac voltages of 1 mV or less. Accordingly, a high-gain meter preamplifier (or peak-probe preamplifier) is very useful. For this purpose, the Radio Shack 277-1008 mini-amplifier will provide a voltage gain of more than 1500 times. The low-level signal is applied to the input of the miniature amplifier, and the amplifier output is fed to an ac voltmeter, or to a peak-reading probe which in turn is connected to a dc voltmeter.

As an illustration of low-level ac voltage measurement in audio circuits, an input of 50 μV to the preamp provides an output of 75 mV or more. An input of 1 mV to the preamp provides an output of 1.5V or more. Note that a 75-mV level is easily measurable with a DVM. The preamp has a volume control whereby the gain can be set at a chosen level, such as 1500 times. *A built-in miniature speaker is included, which makes the preamp a useful audible signal tracer.*

In preamp operation, the ac voltmeter can be connected across the terminals of the miniature built-in speaker, or the output jack for an external speaker may be used. If the output jack is used, a load resistor of 16 ohms will provide silent operation with normal output loading for the mini-amplifier.

AC Voltage Peak-Hold and Dip-Hold Measurements

AC voltage peak-hold and dip-hold measurements can be made in the same manner as explained in Chapter 3 for dc peak and dip measurements. Since series-diode input is employed in these units, the ac input is changed into pulsating dc, and the hold capacitors charge up to the peak value of the pulsating dc voltage.

Demon Hum

Troubleshooters must be alert to the possibility of false ac voltage measurements due to picking up stray-field hum voltage by exposed test leads when connected into high-impedance circuitry. *If the meter reading changes when you grasp a test lead, 60-Hz stray-field hum voltage is entering the measuring circuit.* The remedy is to use shielded input leads to the DVM. (See Figure 5-10.) Similarly, the probe components should be provided with a shielded housing to ensure that false voltage readings will not be obtained due to stray-field pickup.

CHAPTER 6

AC CURRENT TESTS AND MEASUREMENTS

AC CURRENT MEASUREMENT • LOW-LEVEL AC CURRENT MEASUREMENT • PHASE ANGLE MEASUREMENT • DETERMINATION OF CURRENT LEAD OR LAG • CURRENT MEASUREMENT WITH A PEAK-HOLD UNIT • CURRENT MEASUREMENT WITH A DIP-HOLD UNIT • AC CURRENT FLOW WITH DC COMPONENT • AC CURRENT DEMAND OF A POWER SUPPLY • CLAMP-AROUND CURRENT PROBES • MEASUREMENT OF RIPPLE CURRENT • RIPPLE CURRENT FREQUENCY

AC CURRENT MEASUREMENT

AC current values up to 2A rms can be made with the more elaborate service-type DVMs. However most VOMs, TVMs, and DVMs do not provide an ac current function. As a result, the electronic troubleshooter must utilize an external shunt and operate the instrument on its ac voltage function, as shown in Figure 6-1. The basic considerations are similar to those explained for dc current measurement in Chapter 4.

However ac current measurements must also take the ac wave-shape into account. Few service-type DVMs provide true rms indi-

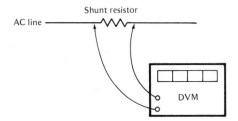

Note: *DVM is operated on its ac voltage function.*

Figure 6-1   Measurement of ac current with a resistive shunt.

cation, so the measured ac current value will be in error unless the current waveform is a pure sine wave. As noted previously, most service-type VOMs, TVMs, and DVMs employ half-wave instrument rectifiers, and they indicate apparent rms values. The ac waveshape can be checked by comparing its apparent rms value with its peak value—*unless the waveshape is a pure sine wave, its apparent rms value will measure more or less than 0.707 of its peak value.*

Note also that the current waveform is not necessarily a pure sine wave although the voltage waveform is a pure sine wave. This is just another way of saying that a constant-voltage ac source with a pure sine waveform will deliver a nonsinusoidal current waveform to a nonlinear load. For example, the power supply in a radio or television receiver draws a highly nonsinusoidal current from a sinusoidal voltage source.

LOW-LEVEL AC CURRENT MEASUREMENT

When low-level ac current values are to be measured, a readable indication can easily be obtained by utilizing a meter preamplifier, as depicted in Figure 6-2. (See Chapter 4 for description of a suitable mini-amplifier.) The preamp should be set for a decimal multiple voltage gain, such as 100 times or 10 times. Note that a resistive voltage divider may be devised for ready calibration of the preamp. For example, if a 1-ohm resistor is connected in series with a 99-ohm resistor, a 100-to-1 voltage division is obtained (see Figure 6-3).

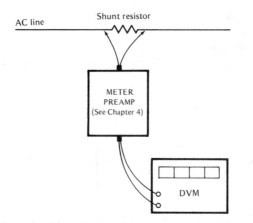

Note: Meter preamp should not be overdriven, or waveform distortion will occur.

Figure 6-2 Low-level ac current measurement.

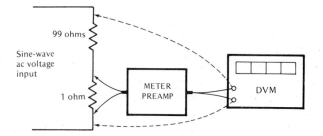

Note: Meter preamplifier is adjusted for a gain of 100 times if the same voltage is indicated at the preamp output as measured at the sine-wave ac voltage input.

Figure 6-3   Calibration of meter preamplifier.

PHASE ANGLE MEASUREMENT

The current in a line is not in phase with the voltage of the line unless the load is purely resistive. In other words, if the load has a capacitive component or an inductive component, the current will be out of phase with the voltage. With reference to Figure 6-4, the phase angle is measured as follows:

1. Connect a meter shunt resistor in series with the line.
2. Measure the line voltage (voltage from 1 to 2).
3. Measure the voltage across the shunt resistor (voltage from 2 to 3).
4. Measure the result of the line voltage and the shunt resistor voltage (voltage from 1 to 3).
5. Combine these three measured voltage values into a triangular diagram, as shown in Figure 6-4(*b*).
6. The angle Θ between 2-1 and 2-3 is the phase angle between the voltage and the current.

Reduction of Line Voltage Component

In various troubleshooting situations the line voltage from 1 to 2 will be very much greater than the resistor shunt voltage from 2 to 3 (Figure 6-4). In such a case, the technician may find it difficult to draw an accurate vector diagram (voltage triangle). To facilitate drawing the diagram, the line-voltage component may be reduced for measurement, as shown in Figure 6-5. Thereby, the measured value from 2 to 1 and from 2 to 3 are more or less equalized. The phase angle is

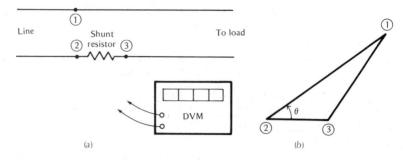

(a) (b)

Note: *Voltage across the shunt resistor is from 2 to 3.*

Voltage across the line is from 1 to 2.

Voltage from 1 to 3 completes the vector diagram.

Angle Θ is the phase angle between current and voltage.

> **Figure 6-4** Phase-angle measurement between current and voltage. (a) Circuit; (b) phase-angle diagram.

calculated in the same manner as if the line voltage had been directly measured.

DETERMINATION OF CURRENT LEAD OR LAG

After the current-voltage phase angle has been measured, the technician may ask whether the current is leading or lagging the voltage. For example, a speaker system will draw a lagging current at one audio frequency, but will draw a leading current at another audio frequency. To determine whether the current is leading or lagging, first measure the current-voltage phase angle as explained in Figure 6-5. Then measure the apparent current-voltage phase angle with a test capacitor in the circuit, as depicted in Figure 6-6.

The voltage from 1 to 2 now leads by virtue of the capacitive reactance. *If a larger phase angle is measured, the load current is lagging the line voltage. If, on the other hand, a smaller phase angle is measured, the load current is leading the line voltage.*

CURRENT MEASUREMENT WITH A PEAK-HOLD UNIT

AC current surges can be measured with a peak-hold unit and a DVM, as shown in Figure 6-7. The shunt resistance value must have a sufficiently high value to provide a readily measurable dc output

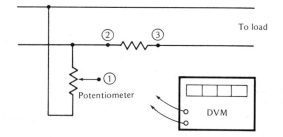

Note: *Use a potentiometer that does not draw a large amount of current from the line. Set the potentiometer so that the voltage from 1 to 2 is more or less the same as the voltage from 2 to 3.*

Calculate the current-voltage phase angle as shown in Figure 6-4.

Figure 6-5   Potentiometer provides reduction of line-voltage component in phase-angle measurement.

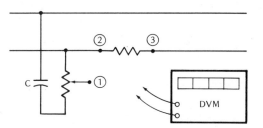

Note: *Capacitor C should have a value that will drop about 25 percent of the line voltage.*

Figure 6-6   Connection of capacitor into measuring circuit for determination of lead or lag.

voltage from the peak-hold unit. The troubleshooter may encounter test situations in which the shunt resistor significantly reduces the load voltage. If this occurs, the measured current value will of course be less than its true value. In such a case, a current correction factor can be applied as follows.

If the shunt resistor drops more than 10 percent of the line voltage, the load voltage and the true load current will also be reduced by the same percentage. This would be a significant reduction in load current. Therefore, the technician should apply a corresponding correction factor to the peak current reading. For example, if the shunt resistor reduces the load voltage by 15 percent, it also reduces the true load current by 15 percent. Thus, the measured

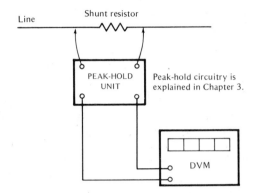

Note: The shunt resistor must have a sufficiently high value to provide a readily measurable output voltage from the peak-hold unit. In turn, a current correction factor may be required (see text).

A 1-ohm shunt resistor will drop 1.414 peak volts for 1 rms ampere of current flow.

A 1-kilohm shunt resistor will drop 1.414 peak volts for 1 rms milliampere of current flow.

Figure 6-7   Current peak-hold arrangement.

peak-current value should be multiplied by 1.18, to obtain the true current value.

Note also that the silicon diode in a peak-hold unit rejects 0.6 volt before it starts conducting. In turn, the measured peak voltage is always 0.6 volt less than the true peak voltage. In most situations, this 0.6 volt loss can be ignored. However, if it is a significant percentage of the peak voltage value, the troubleshooter should add 0.6 volt to the measured peak-voltage value.

CURRENT MEASUREMENT WITH A DIP-HOLD UNIT

AC current dips can be measured with a dip-hold unit and DVM as shown in Figure 6-8. As noted previously, a dip-hold unit cannot distinguish between surge voltages and dip voltages. Therefore the troubleshooter should supplement a dip-hold unit with a peak-hold unit. The same shunt resistor can be used to energize both of the units. Keep the following relations in mind:

1. The load power in watts is equal to the product of the rms current and the rms voltage.
2. If the current leads or lags the voltage, the product of rms current and the rms voltage will be the *apparent* load power.

3. The true load power (real load power) is equal to the apparent load power multiplied by the cosine of the current-voltage phase angle.

4. **Example:** If the output of an audio amplifier works into a 20 μF capacitor, the load has a reactance of 8 ohms at 1 kHz. In turn, an rms voltage of 8 volts across the load produces an rms current flow of 1 ampere. Thus the apparent load power is 8 watts. On the other hand, the true (real) load power is zero, since the current leads the voltage by 90 degrees, and the cosine of 90 degrees is zero.

Use CAT! Punch out cosine values on your pocket calculator. Calculator-aided-troubleshooting can save valuable time on many ac current, voltage, and power calculations.

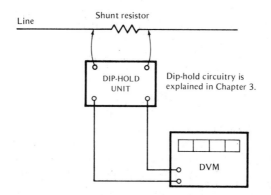

Note: The DVM indicates the peak value of the current waveform. This peak value must be multiplied by 0.707 to obtain the rms current value. (Load power is equal to rms current times rms volts.)

Figure 6-8   Current dip-hold arrangement.

AC CURRENT FLOW WITH DC COMPONENT

When ac current flows through a nonlinear load, it is ac-companied by a dc component. A simple example is shown in Figure 6-9. The diode dimmer in series with the lamp causes the current to flow in half cycles through the load, although the supply voltage has a sine waveform. The ac line current is then accompanied by a dc component. The average value of this dc current component can be measured with the DVM. This is pulsating dc and it has a half-sine waveform, with an average value equal to 0.318 of peak.

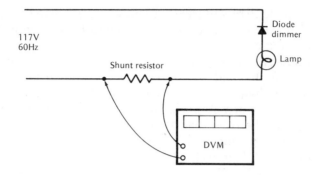

Note: *When a diode dimmer is connected in series with a lamp, the rms value of the current is 0.5 peak, whereas with the diode out of the circuit, the rms value of the current is 0.707 of peak. Since power is proportional to current squared, the lamp might be supposed to operate at half power drain when a diode dimmer is used. However, the filament resistance is nonlinear and this filament resistance will be somewhat less at reduced rms current flow.*

Figure 6-9   Example of a nonlinear load circuit.

The peak value of the current waveform is equal to 3.14 times its average value. As noted in Chapter 5, the true rms value of a half-sine waveform is equal to 0.5 of its peak value. This true rms value can be directly measured with a true rms meter. A conventional ac meter with a half-wave instrument rectifier will indicate an apparent rms value equal to 0.707 of peak—*a reading that is 41.4 percent too high.* Accordingly, the apparent rms reading must be multiplied by 0.707 to obtain the true rms value, in this example.

Note that a conventional ac meter with a half-wave instrument rectifier will read 0.707 of the peak value of the ac current waveform in the foregoing example—*but only when its test leads are suitably polarized.* If its test leads are reversed, the conventional ac meter will read zero. This is just another way of saying that the ac current that flows is accompanied by a dc component that effectively makes the waveform a pulsating dc current.

The ac and dc current components can be separated by connecting a blocking capacitor in series with the ac meter. When this is done, a conventional ac meter with a half-wave instrument rectifier will exhibit turnover in the foregoing example. The waveform that enters the ac meter now has a positive-peak value of 0.682, and a negative-peak value of 0.318, with respect to the pulsating dc peak value of 1.00. In turn, the conventional ac meter will read an apparent rms value of 1.51 for the positive excursion. When its test leads are reversed, the meter will read an apparent rms value of 0.706 for the negative excursion.

Moral: When troubleshooting in nonlinear ac circuits, *know your meters,* and *watch your p's and q's.* Otherwise you will have a classic "tough dog" on your back.

AC CURRENT DEMAND OF A POWER SUPPLY

The foregoing example of nonlinear ac circuitry consisting of a diode dimmer and a lamp is comparatively simple, because its current waveform is a basic standard type. On the other hand, electronic troubleshooters are usually concerned with much more complex types of nonlinear ac circuitry. For example, consider the ac current demand of a half-wave power-supply circuit, such as that depicted in Figure 6-10 (a).

AC current flows through the diode in "pulses" that are approximate sections of half-sine waves. The width of these pulses depends upon the output loading of the power supply. As shown in the diagram, the average value of this current waveform is less than if

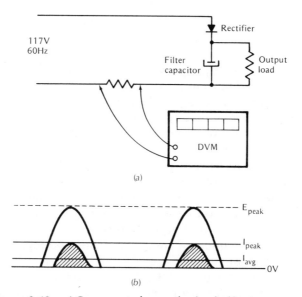

Figure 6-10 AC current demand of a half-wave power supply. (a) Circuit configuration; (b) ac current waveform with moderate output loading.

the pulse width were a full half cycle. Note the following technical pointers:

1. The average value of the pulsating-dc current flow in Figure 6-10 (a) can be measured on the dc-voltage function of the DVM.
2. The peak value of the current waveform can be measured with a peak-hold unit.
3. If the DVM has a true rms function, the actual rms value of the current flow can be measured.
4. If the DVM employs a half-wave instrument rectifier, the apparent rms value of the current flow will be measured.

Troubleshooting of power supplies is often facilitated by ac current measurements on a comparative basis. In other words, electronic service data seldom specifies normal ac current demands for power supplies. However, if a similar item of electronic equipment is available, the technician can make comparative ac current tests. The results of these tests are highly informative.

CLAMP-AROUND CURRENT PROBES

Clamp-around ac current probes are available from various meter manufacturers. These probes are provided with hinged "jaws" that can be clamped around an ac current-carrying conductor. In turn, a proportional ac voltage output is supplied by the probe. This method of ac current measurement speeds up test procedures because the circuit remains intact and does not need to be opened for insertion of a shunt resistor.

A typical ac current probe provides an output of 1 mV for 1 mA of current flow. The probe is accompanied by an amplifier for establishing the measurement calibration level, and also to compensate for the rising high-frequency characteristic of the probe. This ac current probe has a uniform frequency response from 60 Hz to 4 MHz, and can be used in numerous areas of troubleshooting.

AC current probes are also available for power-frequency tests at comparatively high current levels; these probes are not suitable for audio-frequency or video-frequency ac current measurements. Some of these power-frequency current probes are designed for use with a particular VOM. Others are designed for use with any TVM or DVM that has 10-megohm input resistance.

Note that all ac current probes are subject to waveshape form factors, and must be utilized accordingly. This is just another way of saying that an ac current probe does not change a complex waveform into a sine waveform—what goes in must come out insofar as waveshapes are concerned. **Exception:** if you apply a pulsating dc waveform to a clamp-around probe, its dc component is rejected, and an ac current waveform comes out (the average value of the probe output will always be zero).

MEASUREMENT OF RIPPLE CURRENT

Measurement of ripple current in a power-supply output line is often a top-priority troubleshooting requirement. The reason for its importance is the fact that present-day power supplies often have a very low internal impedance (particularly digital-system power supplies). The practical result is that an abnormally large ripple current is accompanied by a very small ripple voltage—this voltage may be almost unmeasurable. On the other hand, an ac current measurement will unmistakably show up the ripple abnormality.

Abnormal ripple current points to excessive current demand by one or more of the branch supply lines. A helpful troubleshooting approach is to disconnect the branch lines one by one, watching the ac current value. If disconnection of a branch line does not result in substantial reduction of the ripple current value, reconnect the branch line and proceed to the next line. The trouble will be found in the branch line that greatly reduces the ripple current level when it is disconnected from the power supply.

In a situation where abnormal ripple current is accompanied by abnormal ripple voltage, look for a fault within the power-supply circuit. For example, the voltage-regulator section is likely to be found defective. A voltage-regulator defect should not be confused with a filter defect—it is easy to make this error because a normally operating voltage regulator has considerable filtering action.

RIPPLE-CURRENT FREQUENCY

Measurement of ripple-current frequency is sometimes of importance in power-supply troubleshooting. For example, the normal ripple-current waveform in a full-wave power supply has a frequency of 120 Hz. On the other hand, any circuit fault that prevents full-wave operation results in a ripple-current frequency of 60 Hz. To measure the ripple waveform frequency, the troubleshooter "bleeds" some sine-wave voltage from a generator into the meter, as explained in Chapter 1, for measurement of pulse width. In turn, the pointer wiggles on the scale when the generator is adjusted to the ripple-current frequency.

CHAPTER 7

AUDIO TROUBLESHOOTING TECHNIQUES

CHECKING AUDIO-AMPLIFIER VOLTAGE GAIN • AVOIDING BIAS-VOLTAGE DRAINOFF • MEASURING STAGE GAIN • TROUBLE SYMPTOMS • SHUT-OFF TEST • DYNAMIC INTERNAL RESISTANCE • DYNAMIC INTERNAL RESISTANCE TEST • COPING WITH NEGATIVE-FEEDBACK LOOPS • BIAS-ON TEST

CHECKING AUDIO-AMPLIFIER VOLTAGE GAIN

Audio-amplifier voltage gain can be checked with a DVM. For example, consider the elementary audio-amplifier configuration shown in Figure 7-1. This arrangement uses an NPN silicon transistor, and typically provides a voltage gain of 120 times. A normally operating stage may have somewhat higher or somewhat lower voltage gain, depending upon the beta value of the particular transistor that is used. *A faulty stage may have little or no voltage gain.*

Observe that both the input terminal and the output terminal of the configuration in Figure 7-1 are above dc ground potential. In other words, the transistor base is biased positive, and this dc bias voltage is present at the input terminal. Also, the 9V battery applies a positive dc voltage to the transistor collector, and this collector dc voltage is present at the output terminal. *These dc voltages that are present at the input and output terminals are a pitfall in the path of the unwary electronic troubleshooter.*

AVOIDING BIAS-VOLTAGE DRAINOFF

To measure the amplifier gain, an audio test signal must be applied, as shown in Figure 7-2. This test signal is coupled to the amplifier transistor by means of a blocking capacitor C_B. Then the

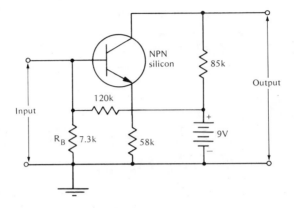

Voltage Gain = 120X

Max. Input Voltage = 0.05V p-p

Max. Output Voltage = 6V p-p

R_B value is comparatively critical. Output waveform has nonlinear distortion.

(The elementary circuit has insufficient negative feedback for high-fidelity operation.)

Figure 7-1 Elementary audio-amplifier configuration.

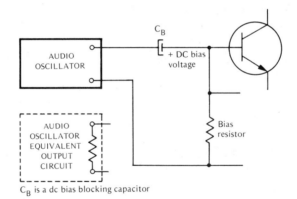

C_B is a dc bias blocking capacitor

Figure 7-2 Blocking capacitor C_B prevents the base-bias voltage from draining off through the audio oscillator.

audio test signal is applied to the transistor base, and the dc base bias voltage remains unchanged. Capacitor C_B should have a value of 50 μF to avoid objectionable attenuation of the audio-oscillator signal, particularly at lower test frequencies. The working-voltage rating of C_B may be quite small because a low-level test signal is used and because the transistor base-bias voltage is less than 1 volt.

Note, however, that C$_B$ is an electrolytic capacitor and should be properly polarized to prevent leakage. In this example the transistor base is biased positively; therefore, the positive lead of the electrolytic capacitor is connected to the base input terminal of the amplifier.

The audio output voltage from the amplifier is measured with a TVM or DVM. A VOM is unsuitable in this application because the amplifier has fairly high output impedance. As a result, *the comparatively low input resistance of a VOM would load the amplifier output circuit objectionably, and a false measurement would be obtained.* If the TVM or DVM has an output function, it can be directly connected at the collector output terminal of the amplifier. On the other hand, if you are using a TVM or DVM that does not have an output function, it will be necessary to connect a blocking capacitor in series with the ac input lead of the TVM or DVM, as shown in Figure 7-3. Otherwise the collector dc voltage would enter the ac meter and upset the instrument-rectifier circuit, and a false measurement would be obtained.

MEASURING STAGE GAIN

The voltage gain of the audio amplifier is measured with an audio test signal applied as shown in Figure 7-2, and with a TVM or DVM connected to the amplifier output terminals, as shown in Figure 7-3. The blocking capacitor in this arrangement has a typical value of 1 μF. This capacitance value is ample, because a TVM or DVM has a high input resistance, such as 10 megohms. Stage gain (in this example) is measured with an audio test signal that has a maximum amplitude of 50 mV peak-to-peak. *If a higher test-signal level were applied to the amplifier input, the resulting amplified waveform would be seriously distorted.*

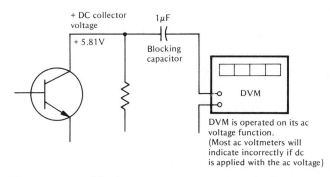

Figure 7-3   A blocking capacitor is connected in series with the ac input lead to the DVM.

Note that some audio oscillators are not designed to provide a low-level output, such as 50 mV p-p. In such a case the troubleshooter must connect a voltage divider between the audio oscillator and the amplifier input terminals, as shown in Figure 7-4. A potentiometer provides convenient adjustment of the voltage-divider output level; it may have a total resistance value of 600 ohms. Before the test procedure is started, it is good practice to set the potentiometer for zero output. Thereby, the possibility of destructive overdrive to the audio amplifier is avoided.

Procedure

The procedure for stage-gain measurement is as follows.

First, a TVM or DVM is connected across the amplifier input terminals; a blocking capacitor is used in series with the ac input lead of the TVM or DVM, as shown in Figure 7-5.

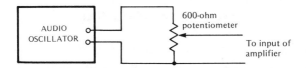

Figure 7-4   A potentiometer can be used to reduce the minimum output level from an audio oscillator.

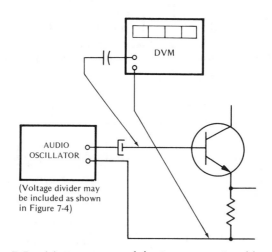

Figure 7-5   Measurement of the input test-signal level to the amplifier.

Second, the test-signal level (in this example) is adjusted to 50 mV peak-to-peak. Note that if your TVM or DVM provides peak-to-peak indication, the millivolt scale may be read directly. On the other hand, if your meter provides rms indication, the scale reading must be multiplied by 2.83 to obtain the equivalent peak-to-peak value. Thus, 17.7 mV rms corresponds to 50 mV peak-to-peak. (Your pocket calculator comes in handy here.)

Third, the TVM or DVM is connected across the amplifier output terminals with a blocking capacitor, as shown in Figure 7-3.

Fourth, the meter scale reading is noted. In this example, normal operation is indicated by an output voltage of approximately 6V p-p, or a corresponding voltage of approximately 2.12V rms.

Note that any audio test frequency in the range from 20 Hz to 20 kHz may be used to check the amplifier gain. It is standard practice to make preliminary audio tests at 1 kHz. Gain tests at low and high frequencies are occasionally of importance in troubleshooting procedures, as explained later in this chapter.

TROUBLE SYMPTOMS

Dead stage: Failure of the ac test signal to appear at the collector of the transistor can be caused by several kinds of defects, which can often be pinpointed by dc voltage measurements. For example, consider the effect of a shorted emitter junction on the dc-voltage distribution, as shown in Figure 7-6. The measurements are:

- Base voltage = 0
- Emitter voltage = 0
- Collector voltage = 9.22V
- Battery voltage = 9.30V

This dc voltage distribution pinpoints the fault to the emitter junction. In other words, *when the base is shorted to the emitter, the transistor cuts off.* Because there is no collector-current flow, the emitter voltage is zero. For the same reason, the collector voltage is the same as the battery voltage. Note that although the collector voltage is the same as the battery voltage, the *measured* value of collector voltage is 9.22V instead of 9.30V. The reason for this difference is that the collector load resistance is 85 kilohms and the DVM input resistance is 10 megohms in this example. This is just another way of saying that the DVM draws a small current and that this current flow drops 0.08V across the collector load resistor. Accordingly, the DVM indicates 99 percent of the actual collector voltage.

The malfunction described above ("dead stage" due to shorted emitter junction) corresponds to a very useful In-Circuit Transistor Quick Test, as explained next.

SHUT-OFF TEST
(IN-CIRCUIT TRANSISTOR QUICK TEST)

A shut-off test provides a very useful quick-check of the transistor condition, as shown in Figure 7-7. The shut-off test is made as follows:

1. A temporary short-circuit is applied between the base and emitter terminals of the transistor.
2. The resulting collector voltage is measured with a TVM or DVM.
3. This collector-voltage reading is compared with the measured terminal voltage of the battery (V_{CC}).
4. If this collector-voltage reading is virtually the same as the V_{CC} voltage, the transistor passes the shut-off test.
5. On the other hand, if this collector-voltage reading is significantly less than the V_{CC} voltage, the transistor does not pass the shut-off test.

In summary, when a temporary short-circuit is applied between the base and emitter terminals of a transistor in a Class-A stage, the collector voltage normally "jumps up" to the V_{CC} value. If there is only a small (or no) increase in collector voltage when this test is made, it indicates that the transistor is defective. As previously noted, a TVM or DVM draws a small current from the collector circuit; for this reason the collector voltage will not quite equal V_{CC} when the shut-off test is made. For example, the collector voltage will measure 99 percent of V_{CC} for a normal transistor in the circuit of Figure 7-1.

Dead stage: Failure of the ac test signal to appear at the collector of the transistor can also be caused by an open collector junction, as shown in Figure 7-8. In this situation, the dc voltage distribution differs from the previous dead-stage example (Figure 7-6). Thus, the dc-voltage measurements in the example of Figure 7-8 are:

- Base voltage = 0.522V
- Emitter voltage = 0.001V
- Collector voltage = 9.22V
- Battery voltage = 9.30V

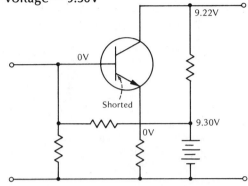

Trouble symptom: Dead stage

Note: Detailed and comprehensive service data for commercial audio amplifiers is published by Howard W. Sams & Co.

Figure 7-6   Example of dc-voltage distribution with shorted emitter junction.

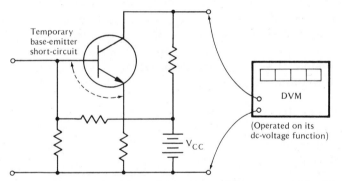

Note: A temporary base-emitter short circuit also provides a very useful quick check in preliminary trouble analysis. Thus, in a multistage audio amplifier, a click will be heard from the speaker when a base-emitter short circuit is applied in a workable stage. On the other hand, no click will be heard from a "dead" stage.

Typical test data:

$V_{CC} = 9.30 V$

Collector voltage measures 3.5 V before short-circuit is applied

Collector voltage measures 9.23 V after short-circuit is applied

(Transistor passes the shut-off test)

Figure 7-7   Test connections for an in-circuit shut-off quick test.

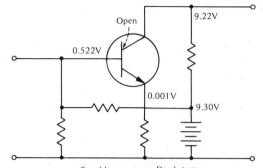

Trouble symptom: Dead stage

Figure 7-8 Example of dc-voltage distribution with open collector junction.

Accordingly, the base voltage of 0.522V clearly distinguishes this dead-stage fault from the previous dead-stage fault. Note also that no change in collector voltage will occur in the Figure 7-6 situation when a shut-off test is made. Similarly, no change in collector voltage will occur in the Figure 7-8 situation when a shut-off test is made.

Weak stage: Subnormal output can be caused by collector-junction leakage, as shown in Figure 7-9. In this example, the collector-junction leakage is 55,000 ohms, and the stage gain is reduced to approximately one-third of its normal value. The resulting dc-voltage distribution differs from the foregoing dead-stage values, as follows:

- Base voltage = 0.556V
- Emitter voltage = 0.006V
- Collector voltage = 1.02V
- Battery voltage = 9.30V

We observe that collector-junction leakage results in an increase in base voltage, and a decrease in collector voltage. Thus, in this example, the base voltage increased from 0.522V to 0.556V, and the collector voltage decreased from 5.81V to 1.02V. *Note also that the transistor does not pass a shut-off test when collector-junction leakage is present.* As an illustration, with a collector-junction leakage of 55,000 ohms, the collector voltage increased from 1.02V to only 3.80V when a shut-off test was made. If the transistor had been in normal condition, the collector voltage would have increased to 9.22V when the shut-off test was made.

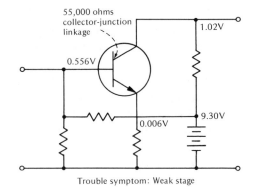

55,000 ohms
collector-junction
linkage

0.556V

1.02V

0.006V

9.30V

Trouble symptom: Weak stage

Figure 7-9   Example of dc-voltage distribution with leaky
collector junction.

DYNAMIC INTERNAL RESISTANCE

In many troubleshooting situations it is helpful to measure the
dynamic internal resistance of a configuration from a chosen device
terminal to ground. As previously explained, the static internal
resistance is measured with a lo-pwr ohmmeter from a chosen device
terminal to ground. Thus, in the configuration of Figure 7-1, the static
internal resistance from emitter to ground is 58 ohms; the static internal resistance from base to ground is 7.3 kilohms; the static internal
resistance from collector to ground is 212.3 kilohms.

When static internal resistance measurements are made in the
configuration of Figure 7-1, the battery is disconnected from the
circuit and the transistor is disabled. Insofar as a lo-pwr ohmmeter is
concerned, a normal transistor "looks like" an open circuit.
Accordingly, if the collector junction were open, a static internal
resistance test could not indicate the fault. Similarly, if the emitter
junction were open, a static internal resistance test could not indicate
the fault.

On the other hand, *when a measurement of dynamic internal
resistance is made, the battery is connected to the circuit, the
transistor is enabled, and the measured resistance value takes the
operating junction resistances into account.* Consequently, a measurement of dynamic internal resistance is much more informative than a
static measurement. Dynamic internal resistance cannot be measured
with any type of ohmmeter; it must be measured in terms of a
voltage/current ratio in accordance with Ohm's law, as explained
next.

DYNAMIC INTERNAL RESISTANCE TEST
(THE MOST COMPREHENSIVE ELECTRONIC CIRCUIT TEST)

Dynamic internal resistance measurements provide the most inclusive information concerning the operating condition of an electronic circuit or network, as shown in Figure 7-10. A dynamic internal resistance measurement is made as follows:

1. The dc voltage is measured from a chosen device terminal to ground.
2. A resistor, such as a potentiometer, is then shunted from the chosen device terminal to ground, and the resulting decrease in terminal voltage is noted.
3. Ohm's law is then applied to calculate the value of dynamic internal resistance.
4. If the calculated value of dynamic internal resistance is normal, it indicates that there is no defect in the corresponding current path.
5. On the other hand, if the calculated value of dynamic internal resistance is abnormal or subnormal, it indicates that there is a defect in the corresponding current path.

In practice, the shunt resistor should have a value that decreases the device terminal voltage by 10 percent, for example. Accordingly, a potentiometer is useful. As an illustration of dynamic internal resistance determination, consider a test from the collector terminal of a transistor to ground in any electronic network. If the terminal voltage initially measures 5 volts and decreases to 4.5 volts when a 45,000 ohm resistor is shunted from the terminal to ground, the dynamic internal resistance is equal to 5000 ohms. In other words, the dynamic internal resistance is equal to 0.5 volt divided by the current drawn through the shunt resistor, or, $0.5/0.0001 = 5000$ ohms. Note that the current drawn through the shunt resistor can be directly measured, although it is usually quicker and easier to calculate E/R. Thus, $4.5/45,000 = 0.0001$.

Let us consider the dynamic internal resistance of the elementary audio amplifier (Figure 7-1) from collector to ground. In normal operation the collector voltage decreased from 5.5 to 4.8V when a 528 kilohm resistor was shunted from the collector terminal to ground. In

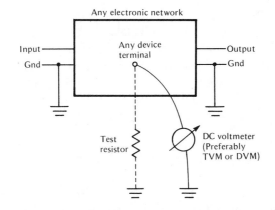

Any electronic network

Input

Gnd

Any device
terminal

Output

Gnd

Test
resistor

DC voltmeter
(Preferably
TVM or DVM)

Figure 7-10   Measurement of dc voltage change for calculation of dynamic internal resistance.

turn, the current drawn by the shunt resistor was equal to $^{4.8}/_{528,000}$, or
9.09 microamperes. Therefore the dynamic internal resistance was
equal to $^{0.7}/_{9.09} \times 10^6$, or 77,000 ohms.

Note that instead of directly calculating the current flow, it is
somewhat quicker and easier to multiply the voltage change by the
value of the shunt resistor and then divide by the value of the
decreased voltage. Thus, $0.7 \times {}^{528,000}/_{4.8} = 77,000$ ohms. Here is a good
application for your pocket calculator. This test showed that the
normal value of dynamic internal resistance from collector ground in
the Figure 7-1 configuration is 77,000 ohms.

*Circuit malfunctions cause changes in the measured values of
dynamic internal resistance.* For example, suppose that there is a
collector leakage resistance of 55 kilohms in the foregoing example.
In this situation the measured value of collector voltage decreases
from 0.928 to 0.898V when a 196 kilohm resistor is shunted from
collector to ground. In turn, the dynamic internal resistance value is
decreased to 6548 ohms.

Again, consider the situation in which the emitter junction is
short-circuited in the configuration of Figure 7-1. The measurements
are as follows:

- Initial collector voltage = 9.22V
- Decreased collector voltage = 8.03V
- Shunt resistance = 528,000 ohms

Thus, the calculated value of dynamic internal resistance is 78,247
ohms. If you protest that it is obvious that the dynamic internal

resistance of the Figure 7-1 circuit from collector to ground with an emitter-junction short-circuit is 85,000 ohms, you are quite correct. How then, are we to evaluate the significance of the foregoing measurements? In this situation, the writer had a hunch that the collector load resistor did not have the value indicated on the diagram, but that it actually had a lower value. Therefore, he disconnected the resistor and measured its value with an ohmmeter. The ohmmeter indicated a value of 78,100 ohms.

Moral: If the measured value of dynamic internal resistance differs from the anticipated value, look for a probable cause. It's in there somewhere.

Weak stage: Subnormal output can also result from human error. For example, if a replacement transistor is connected into the circuit with its emitter and collector leads interchanged, the stage gain will usually be greatly reduced. *Unless the troubleshooter is alert for this kind of malfunction, he can have a tough dog on his hands.* In a typical case history, interchange of the emitter and collector leads resulted in reduction of stage gain to 7 percent of its normal value.

It should not be supposed that the foregoing case history would apply to all transistors. For example, it is possible that you will encounter a symmetrical transistor on occasion. A symmetrical transistor has equal doping concentration in its emitter and collector substances, and it also has equal emitter and collector junction areas. Accordingly, there is no physical distinction between base and emitter characteristics. If the emitter and collector leads of a symmetrical transistor are interchanged, stage gain remains unaffected.

COPING WITH NEGATIVE-FEEDBACK LOOPS

Troubleshooting procedures may be modified in some instances in configurations with negative-feedback loops. Consider, for example, the arrangement shown in Figure 7-11. The 111 kilohm and 19.1 kilohm resistors function both as base-bias resistors and as voltage negative-feedback resistors. In other words, the ac signal at the collector is fed back through the 120 kilohm resistor to the base of the transistor. (There is also a slight amount of current feedback via the 58 ohm emitter resistor.) The negative-feedback loop from collector to base affects stage operation as follows:

1. Voltage gain is reduced to one-half of the value obtained when negative feedback is not used, as in Figure 7-1.

2. The maximum input voltage rating is greater, and the maximum output voltage is less than when negative feedback is omitted.
3. Negative feedback reduces the nonlinear distortion produced by the transistor; however, the elementary circuit in Figure 7-11 has insufficient negative feedback for low-distortion operation.
4. Negative feedback improves operating stability; the circuit in Figure 7-11 has less tendency to drift than the circuit in Figure 7-1.

Stage gain for the Figure 7-11 configuration is measured in the manner previously described for the Figure 7-1 configuration. However, an in-circuit shut-off test cannot be made in the Figure 7-11 arrangement unless an expedient is used. In other words, if the troubleshooter short-circuits the base terminal to the emitter terminal in Figure 7-11, the collector voltage does not increase to the battery

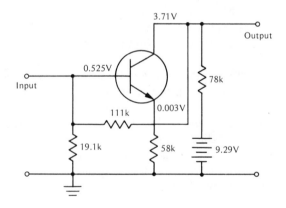

Voltage Gain = 60X

Max. Input Voltage = 0.075V p-p

Max. Output Voltage = 4.5V p-p

Output waveform has nonlinear distortion. (Although some voltage feedback and a small amount of current feedback are provided, the elementary circuit has insufficient negative feedback for high-fidelity operation.)

Figure 7-11   Elementary audio-amplifier configuration with negative feedback from collector to base.

voltage. Instead, the collector voltage increases only to 5.41V. The reason, of course, is that the 111 kilohm resistor draws current from the collector into the base-emitter circuit.

Therefore, troubleshooters sometimes use an expedient in this situation to facilitate a shut-off test. This expedient consists of making a razor cut through the printed-circuit conductor to the 111K resistor. After the turn-off test is completed, the PC conductor is repaired with a drop of solder. Since this expedient complicates the quick test, a bias-on test is often preferred, as follows.

BIAS-ON TEST
(IN-CIRCUIT TRANSISTOR QUICK CHECK)

A bias-on test provides a useful quick check of transistor condition in situations in which a shut-off test is not feasible. A bias-on test is made as shown in Figure 7-12. The procedure is as follows:

1. A temporary "bleeder" resistor is applied between the battery + and base terminals of the transistor.
2. The resulting collector voltage is measured with a TVM or DVM.
3. This collector-voltage reading is compared with the collector-voltage reading that is obtained when the "bleeder" resistor is disconnected.
4. If the collector voltage decreases when the "bleeder" resistor is applied, it indicates that the transistor is workable.
5. On the other hand, if the collector voltage remains unchanged when the "bleeder" resistor is applied, it indicates that the transistor is unworkable.

In the example of Figure 7-12, the collector voltage decreased from 3.71V to 0.10V when the 108 kilohm bleeder resistor was applied. This indicated that the transistor was workable. Note that if a transistor does not pass the bias-on test, the transistor is not necessarily defective—a circuit fault could make the transistor "look bad." For example, a bias-circuit fault could occur.

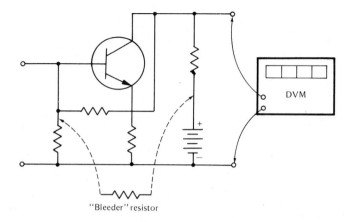

"Bleeder" resistor

Figure 7-12   A 100 kilohm resistor is applied as shown by
the dashed lines to make a bias-on test.

CHAPTER 8

RADIO TROUBLESHOOTING TECHNIQUES

PRELIMINARY TROUBLE ANALYSIS • PRELIMINARY RESISTANCE CHECK-OUT • SIGNAL-VOLTAGE LEVELS • "ELECTROMAGNETIC BATH" QUICK CHECKS • SIGNAL-SUBSTITUTION TESTS • TRANSIENT SIGNAL INJECTION • VOLTAGE GAIN MEASUREMENTS • TRUE SIGNAL-LEVEL MEASUREMENTS • FM OUTPUT FROM AM GENERATOR • SIGNAL-TRACING TESTS • SEC-OND-HARMONIC IF SIGNAL-TRACING PROBE • ISOLATING RESISTORS FOR DVM • CB RADIO TROUBLESHOOTING • FM/CW IDENTIFIER PROBE

PRELIMINARY TROUBLE ANALYSIS

Preliminary trouble analysis of a malfunctioning radio receiver proceeds as follows:

1. When feasible, discuss the onset and nature of the trouble symptoms with the receiver owner or operator.
2. Check the battery voltage (if receiver is battery operated).
3. Open the case or cabinet and try to locate the fault by sight, sound, touch, or smell.
4. Observe the responses to operation of all the receiver controls.
5. If chassis is dusty or dirty, clean it thoroughly—in many cases the trouble will be spotted before the cleaning job is completed.

Quick Checks

In case further trouble analysis is necessary, useful quick checks can be made as follows:

1. *Click Test.* Hold the speaker close to your ear and turn the power switch off and on. A click is normally heard. If there is

117

no click, look for an open circuit in the speaker, output transformer (if used), or their associated connections.

2. *Current Demand.* Disconnect one of the battery clips and insert a milliammeter in series with the battery circuit. There is normally zero current demand when the power switch is turned off. If current does flow, look for a defective switch or a fault in the associated wiring.

3. Next, turn the receiver on and check for normal current demand. For example, a typical pocket radio receiver drew 9 mA with the volume control set to minimum. (A current demand of half this value, or of double this value, would point to a malfunctioning stage.) Excessive current demand can result from accidental reversal of battery polarity. In turn, electrolytic capacitors may be damaged, and transistors may fail catastrophically. Then, when the battery is reconnected in correct polarity, excessive current demand may persist.

4. If the current demand is reasonable, proceed to turn up the volume control. Hold the speaker close to your ear and listen for noise output. If there is very little or no noise, the audio driver stage may be dead, or the detector section may be dead.

5. If the noise output is normal, try tuning the receiver to a local station. Any signal output, no matter how weak, noisy, or distorted, can give helpful clues concerning the trouble location. (See also Chart 8-1.)

Example: A typical pocket radio receiver drew 9 mA from the battery when tuned between stations. On the other hand, when tuned to a local station, the current demand increased to 20 mA. Next, a similar receiver drew approximately 9 mA from the battery, regardless of tuning, and regardless of the volume-control setting. The trouble was tracked down to a shorted tone-control capacitor, which killed the audio signal and produced a dead-receiver trouble symptom.

6. A fairly common cause of a dead-receiver trouble symptom is an inoperative local oscillator. A helpful quick test is to place the dead receiver near another operating receiver. The operating receiver is tuned to a station between 1 MHz and 1.5 MHz. The tuning dial of the dead receiver is thus varied from 545 kHz to 1.45 MHz. If a loud "birdie" (heterodyne squeal) is heard from the operating receiver, the technician concludes that the local oscillator of the dead receiver is workable. On the other hand, if no "birdie" occurs, the trouble will be found in the local-oscillator section of the dead receiver. (See Chart 8-2.)

Chart 8-1

PRELIMINARY RESISTANCE CHECKOUT

Helpful quick-check trouble clues can be picked up on occasion by means of a preliminary resistance checkout. The most important test in the case of a battery radio receiver is the resistance value between the battery-clip terminals. Proceed as follows:

1. Disconnect the battery from the receiver.
2. Turn the power switch on.
3. Measure the resistance between the battery-clip terminals, with ohmmeter polarity the same as normal battery polarity. (Use hi-pwr ohms function.)
4. Compare the measured resistance values on adjacent ohm-meter ranges, and on both AM and FM receiver functions.

Example: A normally operating standard 10-transistor receiver showed the following resistance values:
AM function: On Rx10k range, 12,000 ohms; on Rx1k range, 15,000 ohms; on Rx100 range, 11,000 ohms.
FM function: On Rx10k range, 10,000 ohms; on Rx1k range, 12,000 ohms; on Rx100 range, 3,900 ohms.
Although there is a reasonable tolerance on the battery circuit resistance from one receiver to another, substantially higher or lower measured values would point to a defect in the battery load circuit.
Note that various transistor junctions are conducting in the fore-going test. For a cross-check, repeat the measurement, using the lo-pwr ohms function of the meter. In the foregoing example, the measured resistance value was 50,000 ohms. (A lower or higher value would point to a defect in the battery load circuit.)

The basis of the foregoing test is the electromagnetic coupling that is present when two radio receivers are placed near each other. Other helpful troubleshooting tests based on inter-receiver coupling are explained later in this chapter.

SIGNAL-VOLTAGE LEVELS

Normal signal-voltage levels for a small radio receiver are indicated in Figure 8-1. Note that an input voltage of $^7/_{1,000,000}$ volt normally becomes amplified to a level of 0.7 volt, or, an input level of

Chart 8-2

"ELECTROMAGNETIC BATH" QUICK CHECKS

An "electromagnetic bath" quick check can be very helpful in preliminary trouble analysis when there is no reception, although more or less noise is heard from the speaker when the volume control is turned up.

This quick check consists of connecting a fairly large coil, such as a flyback transformer secondary coil from an old TV receiver, to the output of an AM signal generator, and placing the coil under or behind the receiver under test. The test procedure is as follows:

1. Set the generator for maximum output with 30 percent amplitude modulation.
2. Turn up the volume control in the receiver under test.
3. Tune the generator through the AM broadcast band. If the modulating tone is inaudible, it indicates that the rf input section of the receiver is defective.
4. Tune the generator to the if frequency (455 kHz). If the modulating tone is heard from the speaker in the radio, it is concluded that the if section is operating. If no modulating tone is heard, the trouble will be found in the if, agc, or detector sections.
5. Switch the generator to high-level audio output. If the modulating tone is heard from the speaker in the radio, it is concluded that the af section is operating. If no modulating tone is heard, the trouble will be found in the af section.
6. The same quick-check procedure applies to FM receivers, except that the rf frequency band is from 88 to 108 MHz, and the if frequency is 10.7 MHz.

---

seven microvolts will be stepped up to a level of 700 millivolts. Signals in the microvolt range are called *low-level signals*. It is impractical to directly measure low-level rf signals with service-type test equipment.

On the other hand, it not difficult to measure comparatively high-level af signals, such as the ac voltage across the speaker in Figure 6-1. For example, a VOM, TVM, or DVM can be used to measure the voltage across the speaker, and in turn to calculate the power output from the af section of the receiver. Of course, a sine-wave signal must be employed for this purpose—the receiver must be operated by the output from an AM generator.

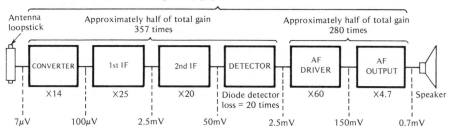

Figure 8-1 Typical rms signal voltages in a small radio receiver.

Note: *These are typical rms signal voltages at successive stages in a small radio receiver. Test signal is from an AM generator with 30 percent modulation. Generator signal is coupled into the antenna loopstick by means of a small coil connected at the end of the generator output cable.*

SIGNAL-SUBSTITUTION TESTS

Electronic troubleshooters often use signal-substitution tests to determine where the signal is being blocked (or attenuated) in the signal channel. A pocket signal injector (noise injector) is often used in preliminary troubleshooting procedures. It consists of a small pulse generator with a repetition rate of approximately 200 kHz. The test signal can be injected at any point in the receiver signal channel, and a "rushing" sound will normally be heard from the speaker. (See also Chart 8-3.)

Note that useful noise-injection tests require that the receiver volume control be adjusted so that little or no noise is audible until the injector tip is touched to a chosen terminal in the receiver signal channel. Otherwise, stray field pickup from the injector will confuse the test. The results of noise-injection tests are not always clear-cut or conclusive—noise-injection tests must often be followed up with AM signal-generator tests.

AM Generator Signal Injection

Signal-substitution tests with an AM generator have several basic advantages over noise-injection tests:

1. The generator provides an adjustable carrier frequency, with a sine waveform, and with an adjustable output level.
2. An AM generator provides a CW output, or an AM output with adjustable modulation percentage.

3. The modulating signal has a typical audio frequency of 1 kHz, with a sine waveform.
4. Sinusoidal test signals permit meaningful voltage measurements with a VOM, TVM, or DVM.
5. The audio modulating signal is available separately for injection into af circuitry.
6. External modulation facilities are provided, whereby the rf carrier can be modulated by any chosen low-frequency signal.

Chart 8-3

TRANSIENT SIGNAL INJECTION

Highly informative trouble clues can often be picked up by means of transient signal injection. As seen below, the base and emitter terminals of a transistor are usually at slightly different voltages.

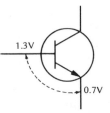

For this reason, a temporary short-circuit applied between the base and emitter terminals brings them suddenly to the same potential and thereby generates a sharp transient pulse. This pulse will normally pass through the receiver circuits and produce a click from the speaker, *unless the stage is very weak or dead*.

Note that a sharp transient pulse has very high frequency harmonics. Therefore, this click test is useful in stages that operate at any frequency, if the receiver volume control is turned up to maximum.

Caution: Be careful not to accidentally short-circuit the collector terminal to the base terminal. This error can result in very high collector-current flow and cause catastrophic damage to the transistor.

With reference to Figure 8-1, a service-type AM signal generator can provide all of the signal levels indicated in the diagram, although some generators do not provide a separate af output signal. An ac voltmeter can be connected across the speaker terminals to measure the output signal level. If the audible output is distracting, an 8-ohm resistor can be connected in lieu of the speaker load. *Preliminary tests are made by injecting signal voltages of suitable frequencies and suitable levels from point to point through the signal channel, "working backwards" from the speaker.*

These preliminary signal-substitution tests serve to indicate whether a test signal can "get through" from the point of injection—although the tests provide little or no data concerning stage gain or signal distortion.

VOLTAGE GAIN MEASUREMENTS

Comparative stage-gain measurements are easier to make than are true signal-level measurements. In any gain test, the agc section of the receiver should be clamped for minimum signal level operation—otherwise, agc action will mask if or rf gain measurements and make the test data meaningless. The agc circuit is clamped with a bias box, as shown in Figure 8-2. Clamp the agc line at its normal no-signal level. In turn, the maximum available gain of each stage will be measured.

Comparative stage gain is measured as follows:

1. With reference to Figure 8-3, an ac voltmeter is connected across the speaker (or across a substitute load resistor).
2. An AM signal generator is connected with its ground lead to the receiver ground (common) line, and with its "hot" lead to a 0.1 μF blocking capacitor.
3. The generator is adjusted to a suitable frequency for the circuit under test, such as 1 kHz, 455 kHz, or 1 MHz.
4. Next, the "hot" output lead from the generator is applied at the output of the stage under test (such as the first if amplifier).
5. Using (in this example) a 455-kHz signal modulated 30 percent, the generator output is adjusted for a reasonable indication on the ac voltmeter, such as 30 mV.
6. Then, the "hot" output lead from the generator is applied at the input of the stage under test, and the increased voltage indication on the ac voltmeter is noted.
7. With reference to Figure 8-1, the troubleshooter will expect to obtain a reading of approximately 0.75 volt in this gain test.

Note that only the output signal level is known in a comparative gain test. In other words, the technician does not know the voltage of the injected signal—he has only verified that the gain of the first if stage is normal (25 times). Thus, the gain of any stage between the first if stage and the speaker might be abnormal or subnormal, but this condition would have to be determined by additional stage-gain tests. Similarly, the test described above provides no data concerning the gain of the converter stage.

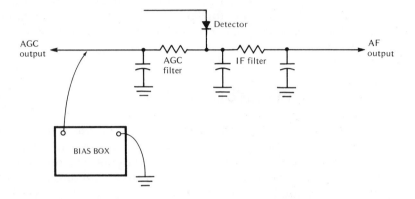

Note: *Observe polarity of agc voltage. Some circuits use positive agc voltage; other circuits use negative agc voltage. This is an example of positive agc.*

Caution: *Some line-operated radio receivers are designed with transformerless power supplies. This type of receiver will have a "hot chassis" unless the power plug is correctly inserted into the power outlet. (Some transformerless receivers have polarized plugs, and do not present the shock hazard of receivers with unpolarized plugs.) A "hot chassis" can also damage test equipment. Therefore, it is good practice to always use a line-isolation transformer when troubleshooting a transformerless radio receiver.*

Figure 8-2 Bias box clamps agc line.

Avoid Overload!

Note carefully that the gain test described above started with a comparatively small output voltage across the speaker. This is an important consideration, because if the technician starts with a substantially higher level, several stages in the receiver would be overloaded in the second step of the test. As a result, the measurement would be in serious error. As a practical rule of thumb, if the output voltage does not exceed its normal rated value, the gain measurement is probably valid.

Nevertheless, there is no *guarantee* that the stage-gain measurement is valid, even if the normal rated output voltage of the receiver is not exceeded. For example, if the af driver stage happens to have incorrect bias voltage, its dynamic range will be reduced, and the transistor will overload before the normal rated output voltage is attained.

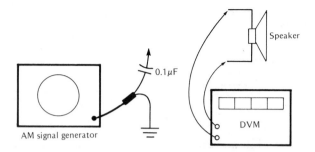

Note: A blocking capacitor is connected in series with the generator output lead to avoid bias drain-off from the circuit under test. The output resistance of the generator is comparatively low, such as 50 or 75 ohms.

Figure 8-3   Test instrument arrangement for gain measurements.

TRUE SIGNAL-LEVEL MEASUREMENTS

Few service-type AM signal generators have a calibrated output attenuator. In turn, the troubleshooter does not know how many microvolts of signal is being applied at a test point. Preliminary troubleshooting procedures do not require this knowledge. On the other hand, follow-up troubleshooting procedures may be greatly hampered unless the technician knows the precise signal level that is being injected into the circuit under test.

Trick of the Trade

A trick of the trade that provides a very useful "microvolt handle" for a service-type AM signal generator is shown in Figure 8-4. The rated input signal levels at various points in the signal channel for the rated output level in a small radio receiver were indicated in Figure 8-1. Thus, a standard pocket radio receiver with a converter stage will normally drive the first if stage at a 100 μV level to produce a 0.7V output level at the speaker.

In turn, the technician can inject a 30 percent modulated if signal at the input of the first if stage, adjust the generator output for a 0.7V reading on the DVM, and then mark the generator output control at the 100 μV setting. (As noted previously, the agc line in the receiver must be clamped at its no-signal voltage value when this type of measurement is made.)

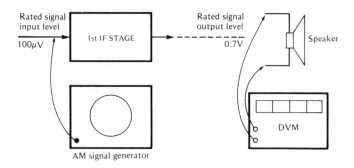

Note: Repeat calibration procedure at each stage in the receiver.

Figure 8-4   Calibration of service-type AM signal generator.

This calibration procedure can then be repeated at each stage in the receiver, so that half a dozen "microvolt handles" will be marked on the generator output control. Although the accuracy of this calibration "dodge" is not equal to that of a lab-type generator rating, it provides a helpful guide in practical troubleshooting procedures.

Note that a test signal should not be directly injected at the input of the converter stage, because the oscillator operation would be seriously disturbed or killed. Therefore an rf test signal is indirectly injected via electromagnetic coupling to the antenna loopstick in the receiver, as shown in Figure 8-5. Any small rf coil, such as an antenna loopstick, may be used. This indirect method of signal injection can

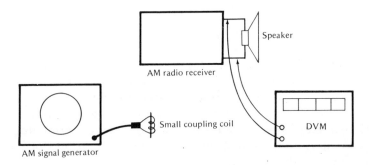

Note: The coupling coil is placed near the receiver and in line with the internal antenna loopstick.

Note: Detailed and comprehensive service data for commercial radio receivers is published by Howard W. Sams & Co.

Figure 8-5   Injection of test signal by electromagnetic coupling.

provide a "microvolt handle" only for a particular type of receiver, and requires precise placement of the coupling coil.

FM OUTPUT FROM AM GENERATOR

Preliminary tests of FM radio receivers can be made with service-type AM signal generators. When an AM generator is tuned to 25 MHz, it will provide some fourth-harmonic output at 100 MHz, for example. If the AM generator is adjusted for a comparatively high percentage of amplitude modulation, more or less incidental frequency modulation will be present in its output. In turn, the modulating tone will normally be heard from the FM receiver when its antenna rod is placed near the generator output cable.

Note that the incidental FM signal from an AM generator is not suitable for signal-injection tests in FM receiver circuitry. The reason for this is that the incidental FM voltage is only a small fraction of the accompanying AM voltage. Consequently, this accompanying AM signal voltage will overload a transistor if injected at its base. *Only an FM signal generator, such as a stereo-FM signal simulator, is suitable for signal-injection tests in FM receiver circuitry.*

SIGNAL-TRACING TESTS

Signal-tracing tests are sometimes preferred in preliminary troubleshooting procedures, chiefly because a signal generator is not required. A DVM can be used to measure signal levels in the detector and audio sections. On the other hand, a service-type DVM does not have sufficient ac-voltage sensitivity to measure if or rf signal levels. (The only exception is the if signal voltage at the detector input *provided that the DVM frequency range extends to 455 kHz.*)

SECOND-HARMONIC IF SIGNAL-TRACING PROBE

IF signal tracing in an AM radio receiver can be accomplished to good advantage with the aid of a second-harmonic if signal-tracing probe, such as depicted in Figure 8-6. This if probe is used in combination with a pocket radio receiver. Its operating features are as follows:

1. The signal-processing portion of the probe consists of a bridge rectifier for the if signal—it operates in this application as a frequency doubler.

2. In turn, the 455-kHz signal input to the bridge circuit is outputted as a 910-kHz rf signal.
3. This 910-kHz rf signal is applied to an antenna loopstick coil which is tuned to 910 kHz.
4. The coil is placed near a pocket radio receiver which is tuned to a frequency of 910 kHz. In turn, the 910-kHz signal is electromagnetically coupled into the receiver.
5. Since the receiver has a high overall gain, even weak if signals become clearly audible after processing by the probe and amplification by the receiver.
6. This probe should be operated on the basis of an AM generator signal coupled into the receiver under test, as depicted in Figure 8-5. The AM generator may be tuned to any frequency within the broadcast band, and the receiver under test must be tuned to the same frequency.

Note that a second-harmonic if signal-tracing probe is used to determine whether signal voltage is present or absent at the point of test. In other words, this probe does not provide true signal-level measurements. Its usefulness is limited to preliminary troubleshooting procedures.

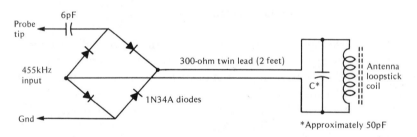

Note: The value of C is chosen to resonate the coil at 910 kHz. Place the coil near a pocket radio receiver. The receiver functions as a probe amplifier and tone indicator. Energize the receiver under test from an AM signal generator.

Figure 8-6   Second-harmonic if signal-tracing probe.

ISOLATING RESISTORS FOR DVM

Both dc-voltage measurements and ac-voltage measurements in radio-receiver circuitry may require the use of a pair of 100-kilohm isolating resistors. One resistor is connected in series with the "hot" meter lead, and the other is connected in series with the gnd or return meter lead. In turn, the resistors become the tips for the meter leads. It is helpful to place each of the isolating resistors inside of a mini-clip. Thereby, test connections can be quickly and easily made in radio circuitry without danger of shorting a base terminal to a collector terminal, for example.

Use of a pair of isolating resistors is often desirable, because an rf stage such as the converter circuit operates with both sides above rf ground potential. In turn, an isolating resistor in both of the meter leads will prevent objectionable disturbance of the circuit under test. As noted previously, substantial circuit disturbance causes incorrect meter readings.

A pair of 100-kilohm isolating resistors will reduce the DVM readout by 2 percent (if the DVM has an input resistance of 10 megohms). In practical service work, this measuring error is usually neglected. However, in critical situations, such as measurement of base-emitter bias voltages, the technician can divide the DVM readout by 0.98 to obtain the true voltage value. (A pocket calculator makes this calculation very easy.)

As a practical sidelight on the desirability of using a pair of isolating resistors with a DVM, we will find that some small AM radios are normally "touchy" and tend to have "hot" circuitry that is "cold" in larger radios. For example, one standard brand of pocket radio receiver, normally generates "birdies," *unless the battery is placed in its compartment, and unless the PC board is in place over the speaker.*

CB RADIO TROUBLESHOOTING

A citizen's-band (CB) radio comprises a receiver and a low-power transmitter. Technically, a CB radio is a transceiver that operates in the 27-MHz band. Preliminary troubleshooting procedures include the following tests:

Dead receiver: When a dead-receiver trouble symptom occurs, check each channel for output. Turn up the volume control; turn the squelch control to minimum. Even if there is no incoming signal, noise output will be heard in normal operation. If there is no noise output, the technician concludes that the channel is dead.

Dead transceiver: If there is no noise output on any channel, it is likely that the entire transceiver is dead. Nevertheless, check out the transmitting function while watching the modulation indicator. If the transmitter is functional, the indicator bulb will vary in brightness while the microphone is being spoken into.

Receiver dead, transmitter OK: When there is no noise output on any channel, and the transmitter operates normally, the trouble will be found in the receiver section between the antenna and the detector.

One dead channel: When there is noise output on all channels but one, there is likely to be a defective quartz crystal in the associated channel circuitry, or possibly a defect in the channel-selector switch.

One operative channel: If there is noise output on one channel only, the trouble is probably in the channel-selector switch.

Squelch control inoperative: An unresponsive squelch control directs suspicion to the automatic volume control (AVC) section. Incorrect AVC voltage is likely to result from a leaky capacitor; all capacitors in the squelch section should be checked.

Weak receiver: If reception is weak, and the antenna has been cleared from suspicion, there is a defective stage in the signal channel. It can be localized by means of signal-injection or signal-tracing tests. When the malfunctioning stage is localized, the faulty component or device can usually be pinpointed by dc voltage and resistance measurements.

Receiver overloads: If the receiver overloads and distorts on strong signals, but provides normal reception on weak signals, the AVC voltage is likely to be off-value. A leaky capacitor is the most probable cause.

Reception intermittent: As previously noted, some intermittents are mechanical, some are thermal, and others are activated by transient voltages. The technician should check the volume control and the squelch control for erratic operation. The channel-selector switch should be checked for defective contacts. Intermittents are sometimes caused by defective insulation on leads, by loose pressure contacts, or by cold-soldered joints. Capacitors may become intermittent; resistors, diodes, or transistors occasionally become intermittent. A quartz crystal may develop a marginal defect.

Weak transmitter: Normal rf power output is approximately 3 watts; the collector dc power is normally 5 watts. If the rf power output is subnormal, a systematic checkout of the transmitter section should be followed. Note that power-type transistors are likely suspects, followed by leaky or open capacitors. A power transistor is likely to be damaged by loss of normal antenna loading.

Transmitter off-frequency: Off-frequency transmission is most likely to result from a defective transmitter quartz crystal. A marginally defective crystal may intermittently jump frequency after operating normally for a longer or shorter period.

Transmitter operates without crystal: When off-frequency operation occurs, and rf power output continues although the quartz crystal has been unplugged from the transmitter, it is probable that the oscillator, driver, or rf power amplifier is self-oscillatory. An open decoupling or bypass capacitor in the transmitter section is the most likely culprit.

Poor modulation: If there is weak or no modulation of the rf carrier, or if there is ample modulation but distorted output, the technician looks for a malfunction in the modulator section. The difficulty is sometimes tracked down to a defective microphone. Substitute a known good microphone. If poor modulation persists, a systematic checkout of components and devices in the modulator section must be made.

FM/CW IDENTIFIER PROBE

When troubleshooting in AM transmitter circuitry, distinction between a modulated carrier and an unmodulated carrier can be made with a demodulator probe, such as previously described. On the other hand, when troubleshooting in FM transceiver circuitry, distinction between a modulated carrier and an unmodulated carrier must be made with an FM/CW identifier probe, such as that depicted in Figure 8-7. This is a simple slope-detector configuration in which coil L is tuned to a frequency slightly to one side of the carrier frequency.

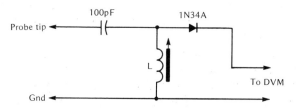

Note: This FM/CW identifier probe outputs a dc voltage if an unmodulated rf carrier is applied. On the other hand, the probe outputs an ac voltage if a frequency-modulated carrier is applied.

Figure 8-7   Basic FM/CW identifier probe configuration.

FM Signal Voltage Levels

FM receivers have typical signal-voltage levels, indicated in Figure 8-8, when normally operating at maximum gain with low-level input. The normal overall signal-voltage gain is greater than for a comparable AM receiver. FM/AM receivers operate on the same basic principles as individual FM and AM receivers; however, some of the circuitry, such as the af section, is common to both the FM and the AM functions.

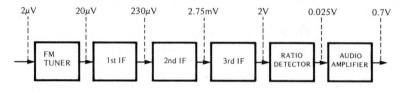

Figure 8-8   Typical signal-voltage levels for an FM receiver operating at maximum gain.

CHAPTER 9

BLACK-AND-WHITE TELEVISION TROUBLESHOOTING

PRELIMINARY TROUBLE ANALYSIS • MEASUREMENT OF POWER-SUPPLY CURRENTS • FOLLOW-UP QUICK TESTS • DC VOLTAGE AND RESISTANCE MEASUREMENTS • AC VOLTAGE MEASUREMENTS • SIGNAL VOLTAGE LEVELS • VERTICAL WAVEFORM IDENTIFIER • HORIZONTAL WAVEFORM IDENTIFIER • TROUBLESHOOTING SELF-OSCILLATION • AGC AND DE-COUPLING CAPACITORS • TROUBLESHOOTING IF REGENERATION • SQUEGGING

PRELIMINARY TROUBLE ANALYSIS

A critical evaluation of the trouble symptoms and control responses for a malfunctioning TV receiver can often save considerable time and effort in following detailed troubleshooting procedures. A preliminary analysis is made as follows:

1. If the situation permits, discuss the onset and nature of the trouble symptoms with the receiver owner or operator.
2. Observe the reponses (if any) to operation of all the receiver controls. Any picture and/or sound reproduction, no matter how weak, noisy, or distorted, can provide helpful clues concerning the type and location of the malfunction.
3. Unless the picture-tube screen is dark, note the snow level on each channel, with the contrast control turned to maximum. Even if there is no picture reproduction, the snow level indicates in a general way where the picture signal is being blocked. For example, a high snow level points to a defective tuner.
4. Open the cabinet and try to locate the fault by sight, sound, touch, or smell. Burned resistors, snapping arcs, loud hum from overloaded transformers, hot electrolytic capacitors, hot power transistors, hot transformers, smoke, odors of burned

133

insulation, and so on, will sometimes direct the trouble-shooter immediately to the faulty component or area.

5. If the chassis is dusty, grimy, or corroded, clean it thoroughly—in many cases, the trouble will be spotted before the cleaning job is completed.

6. Check to verify that all plugs and connectors are properly mated.

Quick Checks

If further trouble analysis is necessary, helpful quick checks can be made as follows:

1. *Power supply input resistance.* Unplug the power cord and measure the input resistance to the power supply with an ohmmeter. **Example:** A Sony receiver normally has a power-supply input resistance of 40 ohms. A substantially lower or higher value would indicate that the power transformer is defective.

2. *Local oscillator operation.* If the snow level is quite high and there is no picture and no sound, the local oscillator may be dead. Make a quick check by bleeding some CW voltage from a signal generator through a 1-kilohm resistor into one of the antenna terminals. If picture and sound are then reproduced as the generator is tuned to the appropriate local-oscillator frequency, the technician concludes that the trouble will be found in the tuner, and that the local oscillator is weak or dead.

3. *Flyback and HV operation.* If there is sound but no raster, the flyback and high-voltage section may be defective, or, the picture tube may be defective. To check, bring an AM pocket radio near the picture-tube screen. If the flyback and HV section is operating, harmonic noise and/or birdies will be heard from the radio at every 15.75 kHz interval on its tuning dial. If there is a loud rushing noise from the radio at all points on the tuning dial when the radio is held at the rear of the TV receiver, there is corona escaping from the HV system.

4. *Evaluation of trouble symptoms.* (See Chart 9-1.)

FOLLOW-UP QUICK TESTS

Follow-up quick tests require that the receiver cabinet be opened. These tests can be very informative, and may save considerable time and effort.

Chart 9-1

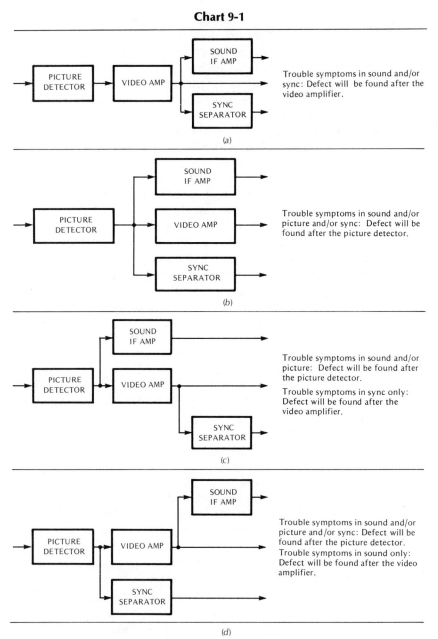

Trouble symptom analysis in basic designs of TV receivers. (a) Sound and sync sections following video amplifier; (b) sound, sync, and video following picture detector; (c) sound and video following picture detector, sync following video; (d) video and sync following picture detector, sound following video.

1. Measure the power-supply output voltage; it should be within ±20 percent of the specified value.
2. Measure the power-supply current drain; it should be within ±20 percent of the value measured in a normally operating receiver of the same type. (See Chart 9-2.) If the supply voltage is low, the current demand will usually be low, also. In case of low supply voltage, look for defective filter capacitors or defective rectifiers. When the current demand is excessive, disconnect the branch load lines one by one until the line is identified that is drawing excessive current.
3. In the event that preliminary tests indicated that the tuner is defective, make a quick test with a tuner subber. This will confirm or clear the initial conclusion.
4. Measure the agc voltage; an agc malfunction can make the rf and/or if sections "look bad." Most agc malfunctions are caused by leaky or open capacitors in the agc system. However, the agc rectifier diode (commonly operated also as the picture-detector diode) can be the culprit.
5. If picture-tube trouble is suspected, first measure the terminal voltages at the picture-tube socket. An off-value bias voltage, for example, can make a good picture tube "look bad."
6. If picture-tube voltages are within tolerance and the screen is dark, make a quick test with a picture-tube test jig.
7. When tracking down malfunctions in the sync section, sync pulses can be injected from a video analyzer. A video analyzer is very useful in follow-up quick tests because it will inject VHF, if, intercarrier-sound, video, and audio signals. The analyzer can substitute for a horizontal or vertical deflection oscillator. It also provides a multi-burst test signal that indicates the overall frequency response of the signal channel at a glance.
8. After a trouble area is located, individual transistors can be checked by transient signal injection, as explained in Chapter 8, and by shut-off or bias-on tests, as described in Chapter 7.

DC VOLTAGE AND RESISTANCE MEASUREMENTS

The majority of detailed troubleshooting procedures involve dc voltage and resistance measurements. DC voltage measurements may be made with signal or without signal, and a comparison of the two values can generally reveal functional data in any nonlinear circuit. Nonlinear circuitry is often characterized by signal-developed bias. Note that the transistor in an oscillator circuit is usually biased by

Chart 9-2

MEASUREMENT OF POWER-SUPPLY CURRENTS

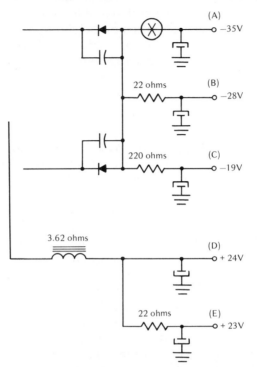

Partial circuit diagram for the power supply in a widely used Motorola TV receiver shows current flowing in five branches.

To measure the current in the (A) branch, open the line at Ⓧ and connect a dc current meter into the circuit.

To measure the current in the (B) branch, measure the voltage across the 22-ohm resistor with a dc voltmeter and calculate the current flow from Ohm's law.

To measure the current in the (C) branch, measure the voltage across the 220-ohm resistor with a dc voltmeter and calculate the current flow from Ohm's law.

To measure the current in the (E) branch, measure the voltage across the 22-ohm resistor with a dc voltmeter and calculate the current flow from Ohm's law.

To measure the current in the (D) branch, measure the dc voltage across the 3.62-ohm choke, calculate the current flow through the choke, and subtract the current that flows in the (E) branch.

Note that an excessive current demand indicates a fault in the associated branch line. Similarly, a subnormal current demand indicates a fault in the associated branch line.

A greatly excessive current demand in one branch line tends to reduce the output voltage in other branch lines, as well as in the fault line. *Check filter capacitors for leakage before assuming that excessive current demand is due to a defect in the receiver circuitry.*

---

signal-developed voltage, and that *reverse bias is a normal condition in all Class-C configurations.*

Specified dc voltage values usually have a tolerance of ±20 percent, except for base-emitter bias voltages. As noted in Chapter 8, an isolation resistor should be used with a DVM whenever there is a hazard of disturbing high-frequency circuit operation. For example, the input capacitance of DVM test leads will detune VHF and if circuits, unless an isolating resistor is utilized. In some receivers this detuning will happen to resonate one stage with the preceding or the following stage. In turn, the high-frequency section will break into self-oscillation and the meter reading will be false.

DC voltage measurements should be made at the line-voltage value specified in the receiver service data. A variable autotransformer or line isolation transformer can be used to adjust the line-voltage value. *A variable isolation transformer is preferred because it eliminates the "hot chassis" problem inherent in the transformerless type of receiver.* A nonisolated "hot chassis" not only shocks the unwary technician but may also damage test equipment. Receiver controls should be set as noted in the receiver service data, and any other specified test conditions should be observed. (See Figure 9-1.)

In-Circuit Resistance Measurements

In-circuit resistance measurements should be made with a lo-pwr ohmmeter, as explained in Chapter 1. Standard service data generally provides resistance charts for TV receivers. These charts specify resistance values from device terminals to ground, measured with a lo-pwr ohmmeter. The tolerance on these values is ±20 percent, unless otherwise noted. Out-of-tolerance resistance measurements may be caused by off-value resistors, by shorted or leaky capacitors, or by defective transistors or diodes.

Note that resistive tolerances are not additive. For example, if two 1000-ohm ±20 percent resistors are connected in series, their

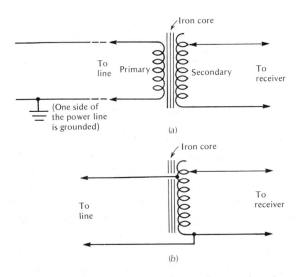

Note: A transformer with separate primary and secondary windings has no conductive path from secondary to primary. On the other hand, an autotransformer has a conductive path from secondary to primary.

If a receiver with a transformerless power supply is energized from (a), its chassis will not be "hot," regardless of how the power plug is inserted into an outlet. On the other hand, if a receiver with a transformerless power supply is energized from (b), its chassis will be "hot" unless the power plug is inserted so that the chassis is connected to the grounded side of the power line.

Figure 9-1 Variable line-voltage devices. (a) Transformer with separate primary and secondary windings; (b) autotransformer with common primary-secondary winding.

resistance value becomes 2000 ohms ±20 percent. Of, if the two 1000-ohm ±20 percent resistors are connected in parallel, their resistance value becomes 500 ohms ±20 percent. Again, if six resistors of various values and ±20 percent tolerance are connected in a series-parallel arrangement, their resultant resistance will have a tolerance of ±20 percent.

Dynamic Internal Resistance

Dynamic internal resistance, explained in Chapter 7, is the most informative resistance measurement from a device terminal to ground. However, dynamic internal resistance (DIR) values are not specified in receiver service data, and this type of resistance measurement must be made on a comparative basis. In other words, a DIR value measured in the receiver under test can only be compared with a corresponding DIR value measured in a normally operating receiver of the same type.

AC VOLTAGE MEASUREMENTS

Few sine-wave voltages are encountered in TV receiver circuitry. On the other hand, many complex voltage waveforms occur, and their peak values and peak-to-peak values must be measured in practical troubleshooting procedures. Peak-to-peak voltages are of basic concern, inasmuch as these values are specified in receiver service data. The peak-to-peak voltages encountered in TV receiver circuitry range from as low as 100 microvolts to 20 kilovolts.

The low-level signal voltages in the input section of a receiver cannot be directly measured in most situations. Medium-level signal voltages and medium-level internally generated waveform voltages can be measured with the aid of the peak-reading and peak-to-peak-reading probes that were previously described. Higher-level internally generated waveform voltages, as exemplified in Figure 9-2, require the use of a comparatively high-voltage probe, as diagrammed.

The complex waveform shown in Figure 9-2(a) is the normal ac waveform at the collector of the horizontal-output transistor in a Sears Model 564.5013 TV receiver. Note that this waveform is actually a dc pulse train; however, it "looks like" an ac waveform to the probe, because the probe circuitry rejects the input dc component. In other words, the probe "sees" only the ac component of the waveform.

Measurement of the peak-to-peak and the peak voltages in this example is straghtforward, *provided that the peak ac voltage inputted to the dc voltmeter does not exceed the range that is being used.* This precaution applies to VOM, TVM, or DVM measurements. To cross-check the possibility of meter overdrive and resulting false readout, switch the meter to the next higher range to determine whether the readout remains the same. If the readout is different, the trouble-shooter concludes that the meter was being overdriven on the lower range.

Measurement of DC Component

Since peak-reading and peak-to-peak-reading probes reject any dc component in a waveform, the troubleshooter must make an

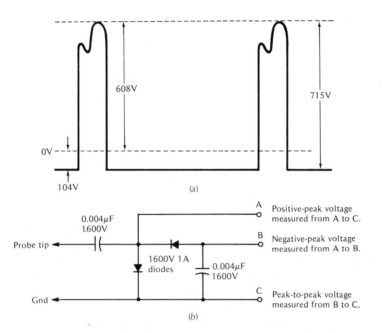

608V

715V

0V

104V

(a)

A — Positive-peak voltage measured from A to C.

0.004μF 1600V

Probe tip

B — Negative-peak voltage measured from A to B.

1600V 1A diodes

0.004μF 1600V

Gnd

C — Peak-to-peak voltage measured from B to C.

(b)

Note: Since the positive-peak voltage is 608V, and the negative-peak voltage is 104V in this example, it would be anticipated that the measured peak-to-peak voltage would be 712V. However the measured peak-to-peak voltage was actually 715V. The 3V difference in the test data is due to experimental error.

If a DVM with a 750V ac range is available, a cross-check will show that the pulse under test is truly a dc pulse. In other words, the DVM (with its half-wave instrument rectifier) reads zero when its test leads are applied in one polarity, and reads 177V when its test leads are applied in the opposite polarity.

Figure 9-2 Example of ac waveform measurement. (a) Pulse voltage at collector of horizontal-output transistor; (b) High voltage +peak/−peak/ peak-to-peak probe for dc voltmeter.

additional measurement in the example of Figure 9-2 to determine whether the waveform under test is an ac waveform, whether it is an ac waveform with a dc component, or whether it is a dc pulse waveform. Therefore the waveform is applied to an RC low-pass filter such as that depicted in Figure 9-3. If the input waveform has a dc component, the filter capacitor will charge up to the dc component voltage.

In this example the capacitor is charged to slightly over 100 volts dc. Since the negative-peak voltage in Figure 9-2(a) measured 104 volts, the conclusion is that the waveform is a dc pulse train. Of course,

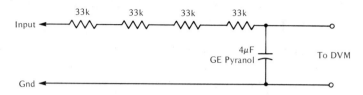
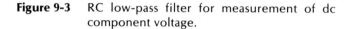

Note: *When resistors are connected in series, their working-voltage ratings are additive.*

Figure 9-3   RC low-pass filter for measurement of dc component voltage.

there is approximately a 4-volt difference between the measured negative-peak value and the measured dc value—a difference of approximately 4 percent. *In practice, this 4-percent discrepancy is regarded as experimental error, and the troubleshooter concludes that the waveform is a dc pulse train.*

Note that it would be difficult or impossible to accurately measure the dc component voltage by applying the pulse waveform directly to the DVM in the foregoing example. First, the positive peaks have a comparatively high voltage; second, the waveform comprises comparatively high ac frequencies for application to the dc circuitry of a service-type DVM. Therefore, the troubleshooter plays it safe by connecting an RC low-pass filter between the waveform source and the DVM.

SIGNAL VOLTAGE LEVELS

It is helpful to note the normal signal voltage levels for a typical TV receiver, indicated in Figure 9-4. Peak-to-peak reading probes can be used with a TVM or DVM to measure the signal voltages in the latter sections of the signal channel, and through the intercarrier-sound section. However the low-level sections are checked by signal substitution, as with a video analyst.

Note that the 4.5-MHz intercarrier sound signal has a normal amplitude of approximately 100 mV at the picture-detector output. It is separated from the video signal by the 4.5-MHz sound takeoff coil (or transformer). In turn, it is amplified through the 4.5-MHz sound-if section, demodulated by the ratio detector, and further amplified to a level of 1.25 V through the audio amplifier.

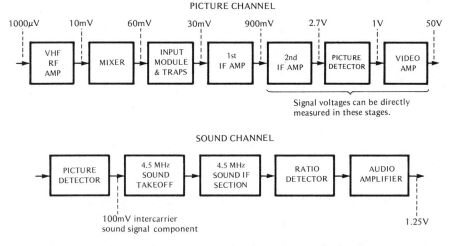

Note: Detailed and comprehensive service data for commercial television receivers is published by Howard W. Sams & Co.

Figure 9-4 Typical signal-voltage levels in a small black-and-white TV receiver.

In the example of Figure 9-4, the overall gain of the picture channel is 50,000 times. The overall gain of the sound channel is 12.5 times. Thus, the overall gain from the receiver input to the audio output is 12,500 times. Note, however, that the sound signal is amplified at 0.1 of the video-signal amplitude from the receiver input to the picture-detector output, where sound-signal takeoff occurs.

VERTICAL WAVEFORM IDENTIFIER

When there is doubt whether a vertical-frequency waveform or a horizontal-frequency waveform is present in the circuit under test, the vertical waveform identifier depicted in Figure 9-5 may be utilized. If a vertical-frequency waveform is applied to the integrating circuit in Figure 9-5(a) a much higher ac voltage will be indicated by a DVM than when the same waveform is applied to the differentiating circuit in Figure 9-5(b).

HORIZONTAL WAVEFORM IDENTIFIER

The differentiating and integrating circuits shown in Figure 9-5 also serve to indicate whether a horizontal-frequency waveform is

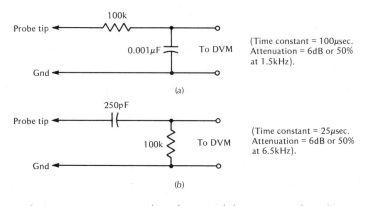

(a)

(b)

Figure 9-5   Horizontal and vertical frequency identifier probes. (a) Integrating circuit passes low frequencies; (b) differentiating circuit passes high frequencies.

present in the circuit under test. If a horizontal-frequency waveform is applied to the differentiating circuit in Figure 9-5(b), a much higher ac voltage will be indicated by a DVM than when the same waveform is applied to the integrating circuit in Figure 9-5(a). This test method is used when the troubleshooter knows that the waveform in the circuit under test may be either a 60-Hz waveform or a 15,750-Hz waveform (but not a mixture of both).

TROUBLESHOOTING SELF-OSCILLATION

Self-oscillation in the if section results in a no-picture and no-sound trouble symptom. Little or no snow is visible in the raster, even with the contrast control advanced to maximum. To verify the preliminary evaluation, measure the dc voltage at the output of the picture detector. A high value, such as 5 or 10 volts, confirms a suspicion of if self-oscillation. There are several possible causes of this malfunction:

1. An agc capacitor may be open. (See Chart 9-3.)
2. A decoupling or bypass capacitor in the if section may be open.
3. The if stages may have been peak-aligned too closely to the same frequency.
4. An incorrect type of replacement transistor may have been used in the if section.

Chart 9-3

AGC AND DECOUPLING CAPACITORS

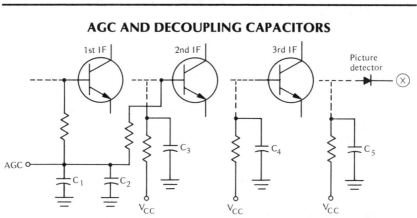

This partial circuit diagram for the if section in a widely used GE TV receiver shows agc capacitors and decoupling capacitors. C_1 and C_2 are agc delay and bypass capacitors. C_3, C_4, and C_5 are collector decoupling capacitors.

DV voltage is measured at ⊗ in the picture-detector output circuit. With no signal present, only a small rectified noise voltage will normally be measured. With signal present, more or less rectified signal voltage will normally be measured, such as 0.5 volt.

Suppose, however, that the three if stages have been accidentally peaked to the same frequency. In such a case, self-oscillation can be anticipated, with the result that a comparatively high dc voltage is measured at the picture-detector output. No signal can then pass through the if amplifier, and the picture-tube raster will be blank, with little or no snow.

The first troubleshooting step is to clamp the agc line at a sufficiently high voltage that oscillation stops.

The second troubleshooting step is to check the if alignment, and to correct the if frequency response, if required. Note that *apparent* misalignment can result from component defects in the if system. *Therefore, all troubleshooting should be completed before alignment is finalized.*

The third troubleshooting step, if required, is a checkout of the agc, bypass, and decoupling capacitors in the if strip.

Note that in a few receiver designs, neutralizing capacitors are included from the base branch to the collector branch of each if transistor. If a neutralizing capacitor becomes open, the associated stage will usually break into self-oscillation.

5. In some receiver designs, shielding in the if strip may have been removed.

To bring if self-oscillation under control, clamp the agc line at a sufficiently high voltage that oscillation stops. Then alignment can be checked and corrected if required. Meaningful voltage measurements can be made, shut-off and bias-on tests can be made, and stage-gain values can be determined. (Self-oscillation nearly always involves the if system, because most of the receiver gain and selectivity is provided by the if amplifier.)

TROUBLESHOOTING IF REGENERATION

Regeneration in the if section results in poor picture quality, and sometimes in distorted sound. Depending upon the adjustment of the fine-tuning control, the picture will have excessive low video-frequency response, with resulting abnormal contrast and loss of picture detail; or, the picture will have excessive high video-frequency reproduction with washed-out background. Circuit ghosts (ringing) accompanies this malfunction in most cases.

The causes of if regeneration are the same as the causes of if self-oscillation—however, the positive feedback that occurs is under the threshold of sustained oscillation. Troubleshooting if regeneration is the same as troubleshooting if self-oscillation. *In all cases, it will be found that the if amplifier has too narrow a bandwidth and a highly peaked frequency response.* Unless the receiver alignment has been tampered with, a faulty component or device will be found to be responsible for the malfunction.

SQUEGGING

Squegging, also called the "Xmas tree" trouble symptom, is another form of self-oscillation that involves the video-amplifier section. It occurs at a much lower frequency than if self-oscillation. It produces odd patterns on the picture tube screen and the sound reproduction may also be distorted. For example, a "motorboating" trouble symptom may occur in the sound output. Sometimes the picture-tube screen may be dark, with only a "motorboating" trouble symptom as a clue to the malfunction.

When squegging occurs, the technician should proceed to *measure the ripple voltage on the* V_{CC} *lines.* In most cases an excessively high ripple voltage will be present. This condition points to power-supply filter trouble—filter capacitors may be found to be defective, or a fault may be found in the regulator section. (Only the larger receivers employ voltage regulator circuitry.) When the ripple voltage has been reduced to a tolerable value, the squegging action will usually be corrected also. However, if the ripple is within tolerance but squegging persists, look for open decoupling capacitors between the video amplifier and the sweep sections of the receiver.

To check for open decoupling capacitors, measure the ac voltage drop across a suspected capacitor with a DVM. If the capacitor is open, a substantial ac voltage will be measured. On the other hand, if the capacitor is in normal condition, a very small or zero ac voltage will be measured across the capacitor. A quick check for a suspected open decoupling or bypass capacitor is to "bridge" it with a good capacitor to see whether the trouble symptom then disappears.

CHAPTER 10

LOW-LEVEL DC VOLTAGE MEASUREMENTS

LOW-LEVEL DC VOLTAGE MEASUREMENTS IN ELECTRONIC TROUBLESHOOT-ING • CONSTRUCTION OF DC PREAMP FOR DC VOLTMETER • NULLING CONSIDERATIONS • SHORT-CIRCUIT LOCALIZATION • SHORT-CIRCUIT TO V_{CC} • CURRENT "WEATHER VANE" • TROUBLESHOOTING "HOT" GROUNDS • "SNEAK VOLTAGES" IN DEFECTIVE GROUND SYSTEMS

LOW-LEVEL DC VOLTAGE MEASUREMENTS IN ELECTRONIC TROUBLESHOOTING

Troubleshooters can speed up and simplify fault identification in various situations by means of low-level dc voltage measurements. *Low-level operation* is not sharply defined—from a practical viewpoint, a voltage value that can be measured only with difficulty or that cannot be measured at all is regarded as a low-level voltage. This practical definition has an extensive gray area, depending upon whether the troubleshooter is using a 20,000 ohms-per-volt meter, a 50,000 ohms-per-volt meter, or a DVM.

For example, if you are using a 20,000 ohms/volt VOM on its 0.25-volt range, the first scale division indicates 5 mV. Any voltage value less than 5 mV will then be regarded as a low-level voltage. However, if you are using a 50,000 ohms/volt VOM with a range doubler on its 0.125-volt range, the first scale division indicates 2.5 mV. Accordingly, any voltage value less than 2.5 mV will be regarded as a low-level voltage.

Again, suppose that you are using a service-type DVM. The smallest change of indication is ±1 mV. Therefore, any voltage value less than 1 mV will be regarded as a low-level voltage. However if you are using a professional-type DVM, a typical indication limit is ±0.1 mV. In turn, any voltage less than 0.1 mV will be regarded as a low-level voltage.

The foregoing examples illustrate a gray area from 0.1 mV to 5 mV. In other words, if a circuit test point has a potential of 1 mV, this is a "low-level test point" for a VOM, whereas it is an "ordinary" test point for a professional-type DVM. As explained next, the trouble-shooter can readily construct a dc preamplifier for a voltmeter, and this preamp will provide a X100 increase of indication sensitivity.

When used with a 20,000 ohms/volt VOM, this preamp provides an indication limit of 0.05 volt, or 50 microvolts. In other words, the preamp makes a 20,000 ohms/volt VOM as sensitive as a typical professional-type DVM. Or, when the preamp is used with a 50,000 ohms/volt VOM with a range doubler, its indication limit becomes 0.025 volt, or 25 microvolts.

CONSTRUCTION OF DC PREAMP
FOR DC VOLTMETER

A suitable voltmeter preamp can be constructed from a Radio Shack Type 741 operational amplifier (op amp). This is an integrated circuit with the package pinout shown in Figure 10-1. It provides a stable dc amplifier with a voltage gain of 100 times when connected as shown in Figure 10-2. The amplifier is powered by two 9V batteries. Note the following points:

1. The voltage gain of the amplifier is equal to the ratio of the feedback resistor value to the input resistor value ($^{1,000,000}\!/_{10,000}$).
2. Pin 2 of the IC is a "virtual ground" due to the substantial amount of negative feedback. This means that the input resistance of the amplifier is 10,000 ohms, in this example.
3. The output impedance (resistance) of the op amp is approximately 75 ohms; in turn, the op amp can be used to drive a VOM regardless of its ohms-per-volt rating.

Consider voltage measurements when the op amp is connected to a service-type DVM that indicates millivolt values. The op amp increases the indication sensitivity by 100 times, so the DVM indicates a voltage level as low as 10 microvolts. In other words, a 1-mV readout corresponds to a 10-μV input voltage to the op amp.

To *zero* the preamp output, short-circuit the input terminals (Figure 10-2). Then make nulling adjustments as follows:

1. Start with the 15k potentiometer set to 10,000 ohms, or slightly more.

2. Adjust the 10k potentiometer to obtain a zero readout on the DVM.
3. If a precise zero readout is not obtained, touch up the adjustment of the 15k potentiometer.

Since these nulling adjustments are quite critical, you may find it helpful to provide vernier adjustments. For example, a 1k poten-

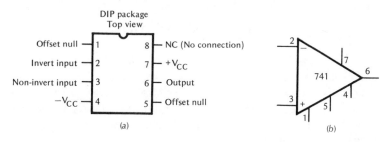

Note: The Type 741 op amp has a maximum available gain of 200,000 times. However, it is always operated with substantial negative feedback, so that the MAG is reduced to 100 times, for example. The op amp is powered by two batteries; the −V_{CC} battery has its positive terminal connected to ground, and the +V_{CC} battery has its negative terminal connected to ground. In most applications the input voltage is applied between the inverting input (pin 2) and ground. Output is taken from pin 6 to ground.

Figure 10-1   Operational amplifier, Type 741. (a) Package pinout; (b) symbol.

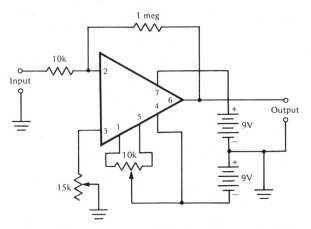

Note: Unless nulling adjustments are provided, the output terminal of the op amp will not rest exactly at zero. This is called the offset voltage. The offset voltage is precisely cancelled out by adjustment of the 15k and 10k potentiometers.

Figure 10-2   Configuration for voltmeter preamplifier.

tiometer may be connected in series with the 15k potentiometer. Similarly, a 1k potentiometer may be connected in series with one end of the 10k potentiometer.

If you have difficulty in nulling out the last vestiges of offset voltage, a final zero adjustment can be made by inserting a millivolt bias box in series with the ground lead of the voltmeter. A millivolt bias box is configured as shown in Figure 10-3.

NULLING CONSIDERATIONS

Troubleshooters do not expect that a DVM will automatically null out precisely on each of its ranges. For example, a professional DVM may normally indicate 0.1 mV (instead of 000.0) when its test leads are short-circuited together. Similarly, a service-type DVM may normally indicate 1 or even 2 mV (instead of .000) when its test leads are short-circuited together.

In other words, a DVM may normally display a small offset voltage when its test leads are short-circuited together. Note that the DVM can be precisely zeroed, in any case, when used with a preamp such as described previously. In other words, the preamp nulling procedure will also cancel out of the DVM offset, if present.

SHORT-CIRCUIT LOCALIZATION

Electronic troubleshooters are often concerned with the localization of a short-circuit. This procedure involves low-level dc voltage measurements. For example, if there is a short-circuit along a printed-circuit (PC) conductor from the collector of a transistor, the circuit will

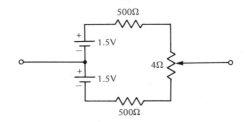

Note: *Although this bias arrangement inserts approximately 250 ohms of resistance in series with the DVM's ground lead, the accuracy of voltage measurement is practically unaffected inasmuch as the input resistance of the DVM is 10 megohms.*

Figure 10-3   Millivolt "bias box" arrangement for final nulling of the preamp output offset.

be dead. In other words, whether the transistor is driven into cutoff or into saturation, its collector circuit will remain dead.

When a transistor is saturated, its collector voltage is typically +0.12V in normal operation. However when there is a short-circuit to ground along the PC conductor from its collector, this voltage drops from +0.12V to zero—*almost*. We will find that this "almost" factor provides a quick and easy method of short-circuit localization, regardless of the circuit's complexity.

Case history: In a simple bistable multivibrator arrangement, a short-circuit in the collector-output circuit resulted in a collector-circuit flow of approximately 10 mA. This current flowed through the PC conductor to ground. *The troubleshooter knew that a short-circuit was present, because a professional-type DVM showed that the output potential was several tenths of a millivolt above zero.*

With reference to Figure 10-4, the troubleshooter started probing along the PC conductor from the collector terminal to points A, B, C, and D. In turn, the DVM readout changed as indicated in the diagram. These low-level voltage readings "finger" the short-circuit in the vicinity of point B.

Next, the troubleshooter started probing along the PC conductor back from the Q output terminal toward the collector terminal of the transistor. *There was no change in DVM reading until the test prod was brought within 1¾ inches of the collector terminal.* When the test prod was moved nearer to the collector terminal, the DVM reading decreased.

The troubleshooter concluded that a short-circuit to ground was located more than 1 inch but less than 2 inches down the PC conductor from the collector terminal (a "ballpark" conclusion).

Next, a close visual inspection showed that a short-circuit within the localized interval had occurred as the result of solder "spread" from an adjoining pad—a replacement component had been soldered to the pad and a short-circuit had accidentally occurred to the adjacent PC conductor. The solder "spread" was camouflaged by a flow of dark varnish.

SHORT-CIRCUIT TO V_CC

Another common fault encountered by the troubleshooter is a short-circuit to V_{CC}. In this situation the PC conductor is at power-supply potential, instead of ground potential. However the basic problem is the same—to find the point along the PC conductor at

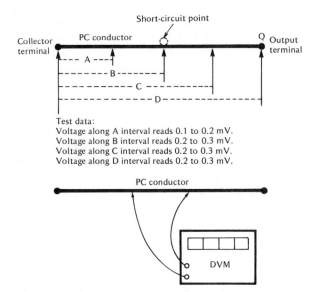

Test data:
Voltage along A interval reads 0.1 to 0.2 mV.
Voltage along B interval reads 0.2 to 0.3 mV.
Voltage along C interval reads 0.2 to 0.3 mV.
Voltage along D interval reads 0.2 to 0.3 mV.

Figure 10-4 Example of short-circuit localization.

which the short-circuit occurs (sometimes the short to V_{CC} occurs at one end of the PC conductor).

The troubleshooter proceeds in the same manner as depicted in Figure 10-4—he probes along the PC conductor, either with a sensitive DVM or with a service-type voltmeter with preamp. When he passes a point beyond which no further voltage increase is measured, the troubleshooter knows that he is close to the short-circuit point.

In the situation wherein the voltage drop along the PC conductor steadily increases (or decreases) from one end of the conductor to another, it is indicated that the short-circuit is at the end of the conductor. (The short-circuit may be at the terminal end: In many cases is will be found that the short-circuit is inside the IC package.)

CURRENT "WEATHER VANE"

Troubleshooters are very familiar with the *voltage distribution* in a circuit. On the other hand, little or no cognizance is usually given to the *current distribution* in a circuit. Nevertheless, this is often a crucial consideration in some electronic troubleshooting procedures. As an illustration, when a short-circuit occurs in a branched circuit (series-parallel circuit), the troubleshooter is likely to fumble and waste considerable time unless he takes the current distribution into account.

First, consider a comparatively simple transistor circuit such as that depicted in Figure 10-5. The *current distribution* is straight-forward—when when Q1 is cut off and Q2 is driven into saturation, current flows up through Q2 and along the PC conductor to Q3. By way of comparison, consider next the widely used branch circuit shown in Figure 10-6. When Q1 is saturated, the total current flows up through Q1 and then branches into currents I_A and I_B.

The current distribution in this case consists of a total current I_T from the collector of Q1 to the branch point, from where it splits into the component currents I_A and I_B (which flow in opposite directions). These current relations are easily checked with a sensitive DVM when its test leads are applied at various points along the PC conductors. *Note that the polarity indication on the DVM display serves as a "current weather vane."*

Next, when a short-circuit occurs in the branch circuitry (at the emitter of Q5 in Figure 10-7), the former current distribution is changed. Whereas the total current flow proceeded from point Y to the branch point in normal operation, now the troubleshooter finds

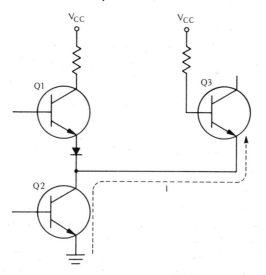

Note: *In this example, when Q1 is cut off and Q2 is saturated, a current I of approximately 1.6 mA flows from ground and through Q2 into Q3. If a short-circuit to ground occurs along this current path, the current value jumps up to about 55 mA.*

Figure 10-5   Example of current flow in a transistor circuit.

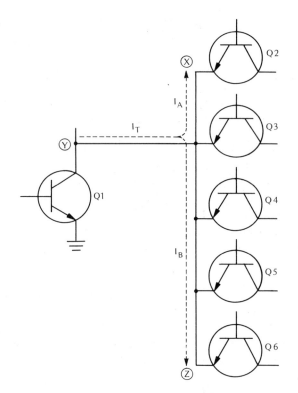

Note: The voltage from the branch circuit to ground is 52 mV, in this example. Current I_A is 1.6 mA. Current I_B is 6.4 mA. The total current I_T is 8 mA. Point X is positive with respect to Y; point Z is positive with respect to Y; point Z is positive with respect to X; point X is negative with respect to Z; point Y is negative with respect to X; point Y is negative with respect to Z. These relations are shown by a sensitive DVM when its test leads are applied between points X and Y, Y and Z, and X and Z.

Figure 10-6 Example of current flow in a branched circuit.

zero current in this section of the PC conductor. His "current weather vane" also shows that the direction of current flow has reversed along the path from point W to V.

When the troubleshooter evaluates this test data with respect to the circuit layout, he recognizes that the short to ground is in the branch conductor from point V to the emitter of Q5. The "current weather vane" is essential in this kind of trouble situation, to avoid a "tough-dog" problem.

This technique is equally useful in case the short-circuit is to V_{CC}, instead of ground. In other words, a short to V_{CC} changes the current distribution in a particular way that points to the fault point, when "current weather vane" test data are taken.

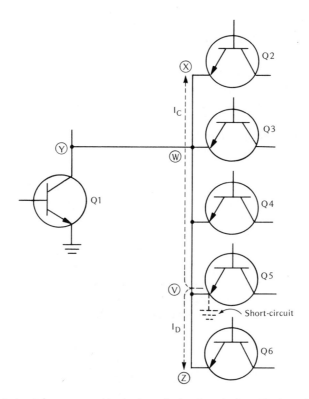

Note: *A short-circuit has occurred in the branch circuit, as indicated. The voltage from the branch circuit to ground is in the order of tenths of millivolts, depending upon the exact spot along the PC conductor at which the DVM test prod is applied. The total current flow is now 55 mA.*

Now there is zero current in the PC conductor from Y to W. Also, the direction of current flow is reversed from W to V.

This change in current distribution from Figure 10-6 is shown by a sensitive DVM when its test leads are applied at various points along the PC conductors.

Figure 10-7   Example of current flow in a branched circuit with a ground fault.

TROUBLESHOOTING "HOT" GROUNDS

Serious "tough dog" troubleshooting problems can be caused by "hot" grounds, which go unsuspected during the course of routine voltage and resistance measurements. *A hot ground is a ground circuit (common circuit) that has abnormally high resistance.* It causes trouble symptoms by coupling signal voltage from one section of the system into another section. For example, hot grounds are frequent causes of malfunction in hi-fi stereo systems.

Here are some useful benchmarks concerning ground-circuit parameters in normal operation:

1. A typical ground-circuit PC conductor on a card has a total length of 18 inches.
2. This ground circuit has a total resistance of 0.85 ohm, or, a resistance of 0.047 ohm per inch.
3. When a current of 65 mA flows through the ground circuit, the total voltage drop is 5.5 mV.

Accordingly, a sensitive DVM is very useful in ground-circuit troubleshooting procedures—the technician passes a known value of dc current through the ground loop, or branch, and measures the millivolt drop across the loop, or branch.

With reference to Figure 10-8, a ground circuit often has various branches. A fault can occur in any section of the circuit. For example, if the millivolt drop is measured from A to B and this value is normal, *it does not follow that the millivolt drop from A to C will be normal.* In other words, a fault can occur in the branch circuit from C to the main AB loop.

Similarly, the millivolt drop from A to D, or from A to E, or from A to F, could be abnormal, although the drop from A to B is normal. Typical causes of ground-circuit faults are:

1. Poor connection of a pigtail to the PC conductor ground.
2. Microscopic crack part-way through the PC conductor.
3. Defective plated-through hole in card.
4. Poor contact of edge connector to card terminals.

Ground-circuit checkouts are easily made by passing a current through the ground circuit from a 1.5V source and a 15-ohm resistor. In turn, the millivolt drop across the ground circuit under test is measured with a sensitive DVM. It is helpful to use a 15-ohm resistor in this test, because a current flow of 100 mA is provided. On occasion, the troubleshooter may make Ohm's-law calculations, and these calculations are simplified by using a decimal value of current.

Caution: When checking out a defective ground system, do not attempt to pass current through a loop by applying a comparatively high test voltage. This practice can result in "making bad devices out of good ones" due to excessive "sneak currents" developed in the defective ground system. (See Figure 10-9.)

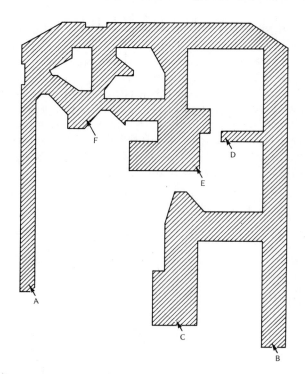

Figure 10-8   Typical ground (common) conductor path on a PC board.

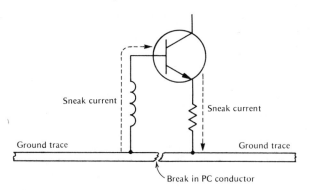

Note: *In this example of a ground-circuit fault, the transistor is protected from sneak-current damage only by its emitter resistor (which has a low value in some configurations). In turn, if a substantial test voltage is applied to the ground loop, the transistor will burn out.*

Figure 10-9   A "sneak current" developed by a ground fault.

"SNEAK VOLTAGES" IN DEFECTIVE GROUND SYSTEMS

"Sneak voltages" accompany "sneak currents" in defective ground systems. This is just another way of saying that comparatively high voltages are often encountered at unexpected points in circuitry associated with a defective ground system. The practical significance of this fact is that, in normal operation, the development of a ground-circuit defect can cause catastrophic failure of several devices or components associated with the ground fault. (An exception to the practical rule of thumb which states that a circuit malfunction is probably caused by a single failed device or component.)

CHAPTER 11

LOW-LEVEL AC VOLTAGE MEASUREMENTS

ACTIVE AUDIO SIGNAL-TRACING PROBE • ACTIVE AM BROADCAST BAND RF SIGNAL-TRACING PROBE • ACTIVE 455-kHz RADIO IF SIGNAL-TRACING PROBE • GAIN OF ACTIVE SIGNAL-TRACING PROBE • TROUBLESHOOTING IF REGENERATION • LOADING EFFECT OF PROBE • POSITIVE FEEDBACK AT FIRST IF STAGE • LOCALIZING OPEN CAPACITORS • CHECK OF ELECTROLY-TIC CAPACITOR RF IMPEDANCE • INCREASING THE INPUT RESISTANCE OF THE ACTIVE PROBE • TROUBLESHOOTING WITHOUT SERVICE DATA

ACTIVE AUDIO SIGNAL-TRACING PROBE

Low-level ac voltage measurements are easily made with active probes that provide substantial voltage gain. The troubleshooter can now make ac voltage measurements in low-level circuitry that was hitherto off limits. An active audio signal-tracing probe for trouble-shooting low-level circuits is depicted in Figure 11-1. This probe provides a gain of 100 times.

Example of application: A DVM with 1-mV sensitivity on its ac-voltage function shows a small voltage indication when connected to the output of a dynamic microphone. (Unless the troubleshooter speaks loudly into the microphone, no voltage indication will be obtained.) However, when the active audio signal-tracing probe is connected between the microphone and the DVM, an indication up to 1V is obtained when the troubleshooter speaks into the microphone.

The foregoing example illustrates the practical utility of the active audio signal-tracing probe. It makes low-level audio-circuit tests feasible—even when a low-level circuit has subnormal gain. The troubleshooter can then quickly localize the point at which a signal becomes weakened (or disappears) in a low-level audio circuit. The probe thus makes "shotgunning" or "Easter-egging" in low-level circuitry unnecessary.

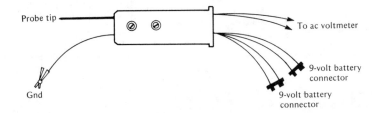

Note: The active audio signal-tracing probe is constructed with the operational amplifier depicted in Figures 10-1 and 10-2. The parts are enclosed in an old-style if transformer housing, and are suspended inside of the housing by their leads. The zeroing potentiometers are mounted on the side of the housing, as shown, so that the amplifier can be easily nulled. In addition to the parts shown in Figure 10-2, this probe includes a 0.22 µF 50WV PC capacitor connected in series with the probe tip inside the housing. (This capacitor functions to block any dc component that may be present, and passes only the ac signal into the op amp.)

Figure 11-1   Active audio signal-tracing probe.

ACTIVE AM BROADCAST BAND
RF SIGNAL-TRACING PROBE

Low-level ac voltage measurements in AM broadcast-band radio circuitry are easily made with an active signal-tracing probe such as that depicted in Figure 11-2. As a result, the troubleshooter can trace weak signals in rf circuitry that were hitherto too weak to detect. Note that since the active probe is tuned to some particular frequency in the AM broadcast band, such as 1 MHz, the radio receiver under test must also be tuned to this frequency. The receiver may be operated from an AM signal generator, or it may be operated from an off-the-air signal.

As a practical note, the troubleshooter may need to shunt a small fixed capacitor across L in Figure 11-2, in order to adjust the probe to a frequency in the lower portion of the AM broadcast band. If desired, a tuning adjustment may be provided in the side of the probe housing, so that the probe frequency can be easily adjusted. Also, the 9V battery should be disconnected from the probe when it is not in use. Note that you may encounter probe instability when testing across a tuned circuit that has a very high Q value. When instability occurs, the meter will indicate a voltage although the circuit under test is "dead." To eliminate probe instability, connect a resistor of suitable value in series with the probe tip, as depicted in Figure 11-5. Although probe instability is rarely encountered, this possibility should be kept in mind.

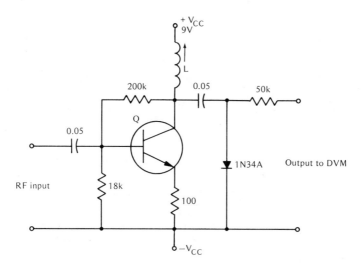

Note: *Transistor Q may be a Radio Shack 276-1603, or equivalent. L is an AM broadcast-band inductor. The 18k bias resistor from base to common lead may need to be changed to a slightly lower or slightly higher value for optimum bias voltage. L may be adjusted to any desired frequency in the AM broadcast band, such as 1 MHz. The active probe develops maximum gain only when the rf input signal has the same frequency to which L is adjusted.*

Note: *The active signal-tracing probe consists of an rf-amplifier stage driving a peak detector. It is much more effective than a peak detector probe followed by an audio amplifier because the detector diode is operated far more efficiently. In other words, a diode has poor rectification efficiency at low signal levels. In turn, this limitation in diode operation is avoided by increasing the rf signal level before detection.*

Figure 11-2 Active AM broadcast rf signal-tracing probe.

ACTIVE 455-kHz RADIO IF SIGNAL-TRACING PROBE

Low-level 455-kHz if voltage measurements in AM broadcast radio circuitry can be made with the same active-probe configuration shown in Figure 11-2, provided that a 455-kHz if inductor is substituted for L. The inductor should be tuned to precisely 455 kHz. The troubleshooter then can signal-trace very weak signals through the if section of an AM broadcast radio.

Another very useful application of an active signal-tracing probe is to check the attenuator calibration of a signal generator. In other words, if the rf output cable from the generator is connected to the input of the active probe, the DVM will indicate the relative voltages of very low-level signals. Thus, the troubleshooter can easily determine whether the X100 and X1000 steps of the attenuator, for example, actually have a 10-to-1 ratio.

Caution: The active probe must not be overloaded, or incorrect conclusions will be drawn. In this example, overload starts when the DVM reading exceeds 8.3 volts. Overload shows up as a limiting action—the DVM reading does not continue to increase as the input signal level is increased.

GAIN OF ACTIVE SIGNAL-TRACING PROBE

The active signal-tracing probe depicted in Figure 11-2 has a typical voltage gain of 65 times. To adjust the gain to a round number, such as 50 times, the value of the emitter resistor may be increased somewhat. On the other hand, if the emitter resistor is bypassed, the gain will be increased.

Example: The 100-ohm emitter resistor in the configuration of Figure 11-2 was bypassed by a 0.05 μF capacitor. In turn, the probe gain increased to over 100 times. By "trimming" the value of the emitter resistor, the probe gain can be adjusted to precisely 100 times.

Waveform Error

Troubleshooters should recognize that although an active probe is adjusted for a precise voltage gain, such as X100, that measurement accuracy will nevertheless be impaired by waveform error. These considerations are as follows:

1. Service-type AM signal generators often have considerable waveform distortion. The amount and nature of this distortion may change rapidly as the generator is tuned from one frequency to another.

2. The amplifier in the active probe rejects most of the harmonics that are present in a distorted input waveform. The output waveform from the probe amplifier approximates a pure sine wave, regardless of the input waveshape. (This is due to high Q of the collector-load inductor, and its "flywheel" effect.)

3. Generator output attenuators are customarily calibrated in rms values of sine-wave voltages. On the other hand, the DVM in an active probe arrangement such as that depicted in Figure 11-2 indicates peak-voltage values. The bottom line is: If you are using a service-type AM signal generator that has a poor output waveform, do not expect to make accurate measurements with the active signal-tracing probe.

TROUBLESHOOTING IF REGENERATION

An active signal-tracing probe is very useful in tracking down if regeneration. As shown in Figure 11-3, there are three common positive-feedback loops to be considered:

1. Positive feedback from the output to the input of the second if stage.
2. Positive feedback from the output of the second if stage to the input of the first if stage.
3. Positive feedback from the output of the second if stage to the input of the mixer stage. The if signal level is maximum at the output of the last if stage. As a result, positive feedback loops are more likely to cause objectionable distortion when the positive feedback starts from the output of the last if stage.

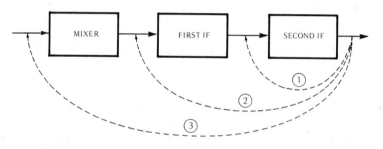

Note: IF regeneration may be caused by incorrect alignment, as when stages that are normally stagger-tuned are peaked to the same frequency. IF regeneration can also be caused by replacement of a transistor with a nonstandard type that has considerably higher beta value. IF regeneration is often caused by open bypass or decoupling capacitors in the if circuitry—an open AVC capacitor may be the culprit. IF regeneration is also caused by a "hot" ground due to a cracked PC conductor or a poor connection.

Figure 11-3   Three common positive-feedback loops.

Regeneration (positive feedback) causes trouble symptoms such as poor sound quality (sideband cutting), critical tuning, and "birdies" when the if circuitry is on the verge of breaking into self-oscillation.

With reference to Figure 11-4, the troubleshooter checks for possible positive-feedback to the input of the mixer as follows:

1. An active signal-tracing probe and DVM are connected at the input of the mixer, as shown in the diagram.
2. The receiver is energized from an AM signal generator in order to maintain a steady if signal level.

3. The if signal level is noted on the DVM display. Then the troubleshooter shunts a 0.05 μF capacitor across the output terminals of the second if stage.

4. If the DVM reading decreases, the troubleshooter knows that there is a positive-feedback loop from the output of the second if stage to the input of the mixer. On the other hand, if the DVM reading remains unchanged, the troubleshooter knows that the positive-feedback loop does not include the second if and mixer stages.

Next, with reference to Figure 11-4, the possibility of positive feedback from the output of the first if stage to the input of the mixer is checked out as follows:

1. An active if signal-tracing probe and DVM are connected at the input of the mixer, as shown in the diagram.
2. The receiver is energized from an AM signal generator.
3. The if signal level is noted on the DVM display. Then the troubleshooter shunts a 0.05 μF capacitor across the output terminals of the first if stage.
4. If the DVM reading decreases, the troubleshooter knows that there is a positive-feedback loop from the output of the first if stage to the input of the mixer. On the other hand, if the DVM reading remains unchanged, the troubleshooter knows that the positive-feedback loop does not include the first if and mixer stages.

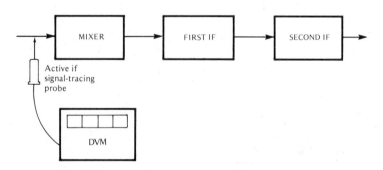

Note: *Whenever positive feedback is present from an if stage to the input of the mixer stage, the DVM will indicate the value of the normal if voltage at the test point, plus the value of the positive-feedback voltage. If the positive-feedback loop is broken by some means, the DVM reading will decrease.*

Figure 11-4   Basic test setup for checking a regenerative if section.

LOADING EFFECT OF PROBE

The active signal-tracing probe has more or less loading effect on the circuit under test. Thus, if the probe is applied at a very high-impedance circuit point, the stage operation may be "killed." As shown in Figure 11-5, there are two ways to minimize the loading effect of the probe. First, the troubleshooter may note a low-impedance signal point in the circuit under test—such as an unbypassed emitter resistor. The probe may be applied across the resistor with very little resulting circuit disturbance.

The second way to minimize the loading effect of an active probe is to connect an isolation resistor in series with the probe tip, as shown in the diagram. *This resistor should have as high a resistance value as possible, without excessive reduction of signal strength.* As long as a substantial value of signal voltage is indicated by the DVM, comparative measurements, as in regeneration checkouts, can be made as effectively as before.

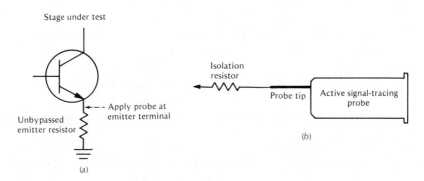

(a)

(b)

Note: The loading effect (circuit disturbance) is much less when the probe is applied at a low-impedance point in the circuit, such as at the emitter of the transistor. An isolation resistor of suitable value reduces the loading effect of the probe. However, there is a tradeoff in terms of reduced signal amplitude.

Figure 11-5 Minimizing the loading effect of the active probe. (a) Apply probe across an unbypassed emitter resistor; (b) use an isolating resistor in series with the probe.

POSITIVE FEEDBACK AT FIRST IF STAGE

In case no evidence of positive feedback is found at the input to the mixer, the troubleshooter should proceed to check next at the input to the first if stage, as shown in Figure 11-6. In other words, a

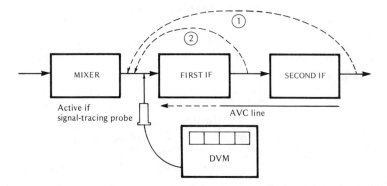

Note: *Troubleshooters will encounter positive-feedback situations in which the feedback takes place via the AVC line. Therefore, it is important to note the result of bypassing the AVC line to ground, as well as shunting a capacitor across the output of the first if or second if stage.*

Figure 11-6   Positive-feedback checkout at input of first if stage.

positive-feedback loop sometimes extends only to the first if stage, and does not include the mixer stage. Test procedure is as follows:

1. An active signal-tracing probe and DVM are connnected at the input of the first if stage, as shown in the diagram.
2. The receiver is energized from an AM signal generator, as before.
3. This if signal level is noted on the DVM display. Then the troubleshooter shunts a 0.05 μF capacitor across the output terminals of the first if stage, and across the output of the second if stage, in turn. Finally, he shunts the capacitor from the AVC line to ground.
4. If the DVM reading decreases in one or more of the foregoing tests, the troubleshooter knows that there is a positive-feedback loop into the first if stage. On the other hand, if the DVM reading remains unchanged, the troubleshooter knows that the positive-feedback loop is localized to a single stage.

LOCALIZING OPEN CAPACITORS

Open capacitors in low-level rf circuitry can cause various trouble symptoms. For example, with reference to Figure 11-7, C1 bypasses R1 and R2 in the base circuit of Q1. If C1 becomes open, the gain of the

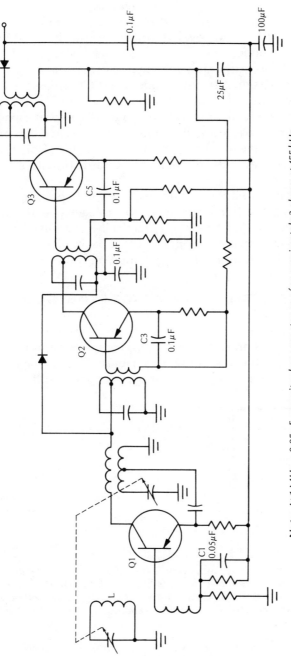

Note: At 1 MHz, a 0.05 μF capacitor has a reactance of approximately 3 ohms; at 455 kHz, it has a reactance of about 7 ohms. In turn, the impedance from the lower end of L to ground is not greater than 7 ohms in normal operation. However, if C1 becomes open, the resistance to ground is then 860 ohms—an increase of over 100 times.

Figure 11-7 High-frequency circuitry in a small radio receiver.

converter stage is reduced. In normal operation, the voltage drop across C1 is virtually zero (rf voltage drop). However, if C1 is open, a large proportion of the low-level rf voltage in the base circuit is dropped across R2 and R1.

A quick check for a suspected open bypass capacitor can be made by applying the active signal-tracing probe across C1. Unless the DVM reads practically zero, the troubleshooter recognizes that the capacitor is open. The same observation applies to a bypass capacitor such as C4. Similarly, C6 is normally "cold." Note also that capacitors C3 and C5 normally have practically zero rf voltage drop from end to end. On the other hand, both ends of C3 and C5 are normally above rf ground potential.

Thus the troubleshooter can quickly check out bypass capacitors with the use of an active signal-tracing probe. In each case the active probe is applied *across the capacitor*—practically zero voltage will be measured, unless the capacitor is open. If the capacitor happens to be open, a substantial indication will be obtained on the DVM.

CHECK OF ELECTROLYTIC CAPACITOR RF IMPEDANCE

Although it might be supposed that the rf impedance of an electrolytic capacitor is practically zero, this is not necessarily the case. For example, when the test shown in Figure 11-8 is made, the troubleshooter generally measures an appreciable voltage drop

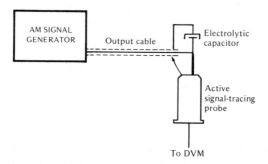

Note: The design of an electrolytic capacitor is such that it has considerably more residual inductance than solid-dielectric capacitors. An electrolytic capacitor has a much higher power factor than a solid-dielectric capacitor. This is just another way of saying that the rf impedance of an electrolytic capacitor is greater than that of a solid-dielectric capacitor.

Figure 11-8   Check of electrolytic capacitor impedance at rf.

across the electrolytic capacitor. Next, if a 0.22 μF capacitor is shunted across the electrolytic capacitor, the voltage drop typically decreases to a small fraction of its former value. This demonstration shows that an electrolytic capacitor should be supplemented with another capacitor, considerably smaller in value, in order to minimize its rf impedance.

Active Signal-Tracing Probe
With Higher Performance

Higher performance is provided by an active rf signal-tracing probe with a slightly elaborated configuration. With reference to Figure 11-2, the probe sensitivity can be doubled by the addition of a diode and capacitor, as shown in Figure 11-9. In other words, peak-to-peak rectification is provided, instead of peak rectification. As previously noted, the probe can be calibrated by varying the value of the emitter resistor somewhat. Also, the gain can be increased, if desired, by bypassing the emitter resistor.

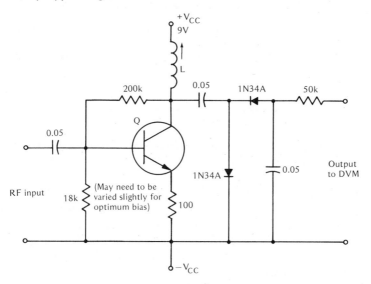

Note: *This arrangement is basically the same as the active-probe configuration depicted in Figure 11-2, with the exception that a peak-to-peak rectifier circuit is used, instead of a peak rectifier circuit. The additional diode and capacitor effectively provide double sensitivity for the active probe.*

Figure 11-9　Elaborated active-probe configuration provides double sensitivity.

INCREASING THE INPUT RESISTANCE
OF THE ACTIVE PROBE

Troubleshooters sometimes need an active probe with higher input resistance than is provided by the arrangement shown in Figure 11-5(b). Higher input resistance can be obtained by doubling the value of the isolation resistor, and using a peak-to-peak active probe instead of a peak active probe. Thereby, circuit loading can be reduced when signal-tracing in rf or if circuitry.

Voltage Magnification

The maximum output level of the active probe shown in Figure 11-9 is approximately 20V p-p. Although this might seem to be an "impossible" output level because V_{CC} is only 9V, this 11-volt increase over V_{CC} is provided by voltage magnification in the collector-load circuit. In other words, L operates both as a parallel-resonant circuit and as a series-resonant circuit. L is series resonant with the effective capacitance at the collector terminal. The rf output voltage developed by the transistor is then magnified by the Q value of the series-resonant circuit. In this example, the voltage magnification is more than double the rf voltage outputted by the transistor alone.

TROUBLESHOOTING WITHOUT SERVICE DATA

Troubleshooting without service data is a "troubleshooter's curse." However the task can be considerably eased by first mapping out the circuitry with tuned signal-tracing probes. (An active rf signal-tracing probe is tuned to 1 MHz, for example; an active broadcast-if signal-tracing probe is tuned to 455 kHz.) Since a tuned rf probe does not respond to an if signal, and a tuned if probe does not respond to an rf signal, the troubleshooter can quickly map out an unknown receiver (excepting stages that might be completely dead).

The same mapping-out technique can be used in troubleshooting FM receivers without service data. In other words, the collector load inductor in an active FM signal-tracing probe is chosen to resonate at 10.7 MHz (the standard FM if frequency). In FM/AM receivers, the troubleshooter can quickly separate the FM and AM stages by means of tests with tuned signal-tracing probes (excepting stages that might be completely dead).

Audio stages in either AM or FM receivers can be quickly identified by means of an active audio signal-tracing probe. In other words, the op amp in an active audio probe functions as a low-pass filter in conjunction with the ac-voltage function of the DVM. Thus, an active audio signal-tracing probe will not respond to rf or if signal voltages.

SECTION II

DIGITAL TROUBLESHOOTING TECHNIQUES

CHAPTER 12

FAMILIARIZATION

EXAMPLE OF LATCH TROUBLESHOOTING WITH A VOM • NOR-GATE D-TYPE LATCH OPERATION AND TROUBLESHOOTING • FLOATING INPUT TERMINAL "LOOKS" LOGIC-HIGH • FAMILIARIZATION EXPERIMENTAL PROJECT • CIRCUIT ACTION OF NOR GATE • D LATCH SOURCE RE-SISTANCE REQUIREMENT

EXAMPLE OF LATCH TROUBLESHOOTING
WITH A VOM

A latch is a bistable multivibrator. It is used for temporary storage of binary data in a digital system. This is just another way of saying that a latch is the simplest form of digital *memory*. Binary data consists of logic-high (1) and logic-low (0) states. In the latch described in this chapter, a logic-high state is a voltage greater than +2.4 volts; a logic-low state is a voltage less than +0.4 volt. This is called a *positive-logic* mode.

The troubleshooter should clearly distinguish between a *latch* and a *flip-flop*. A latch is an asynchronous device—it is unclocked and can be operated at any time. On the other hand, a flip-flop is a synchronous device—it is clocked and operates in step with clock pulses. In other words, a flip-flop is a clocked (synchronized) bistable multivibrator, and is somewhat more elaborate than a latch. As explained in greater detail subsequently, a latch is *loaded* by inputting a logic-high (1) level, for example. The latch then stores this binary digit until another binary digit is inputted—a logic-low (0) level, for example. Thereupon, the latch normally outputs or *unloads* the 1 as it *loads* the 0.

If the latch fails to load and to unload when binary digits (bits) are applied to its input terminal, the fault may be tracked down to either the integrated circuit that contains the latch, or to a defect in a branch circuit. In any case, the troubleshooter can localize the malfunction by means of dc-voltage measurements with a VOM (or DVM), as explained subsequently. (Always measure V$_{CC}$ first!)

NOR-GATE D-TYPE LATCH OPERATION
AND TROUBLESHOOTING

The D-type latch configured from NOR gates is in wide use; a 2-input NOR gate has a symbol, truth table, and logic equation as shown in Figure 12-1. Note that the NOR gate's output will change state in response to an input signal of 2.0 volts; an output stage change from logic-low to logic-high consists of an output-voltage rise from approximately 0.09 volt to 3.75 volts. *Thus, the NOR gate provides voltage amplification.*

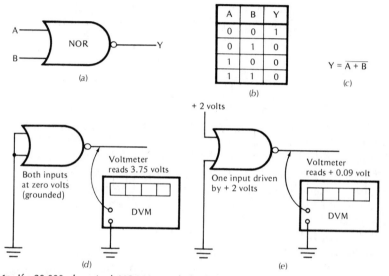

A	B	Y
0	0	1
0	1	0
1	0	0
1	1	0

$Y = \overline{A + B}$

(a)

(b)

(c)

+ 2 volts

Both inputs at zero volts (grounded)

Voltmeter reads 3.75 volts

DVM

One input driven by + 2 volts

Voltmeter reads + 0.09 volt

DVM

(d)

(e)

Note: If a 20,000 ohms/volt VOM is used, the logic-low level must be measured on the 50 μA (0.25V) range of the meter. Always check first with the 5V range of the meter—in case the point under test might be logic-high.

Figure 12-1   NOR gate data. (a) Symbol for 2-input NOR gate; (b) truth table; (c) logic equation; (d) when both inputs are at zero volts, the output is normally at 3.75 volts; (e) when one input is at +2 volts, the output is normally at 0.09 volt.

In turn, a pair of cross-connected NOR gates operates as a bistable multivibrator, or latch, as shown in Figure 12-2. This is the basic RS (reset-set) latch; its circuit action is summarized in the truth table shown in (b). *Observe that a NOR gate has a V_{CC} connection and a gnd connection, as depicted in (d).* The V_{CC} and gnd connections are

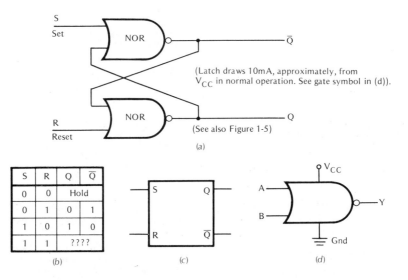

Note: The logic diagram in (a) shows that a reset-set (RS) latch is a bistable multivibrator with inputs S and R, and with outputs Q and $\bar{Q}$. The truth table in (b) shows that if both inputs are logic-low (grounded), the outputs do not change state—they simply remain in their previous states. Next, if S is logic-low and R is logic-high (connected to V_{CC}), Q will go logic-low, and $\bar{Q}$ will go logic-high. Then, if S is driven logic-high and R is driven logic-low, Q will go logic-high, and $\bar{Q}$ will go logic-low. **Caution:** If S and R are both driven logic-high, the output states will be unpredictable—it is "forbidden" to drive S and R logic-high at the same time.

Figure 12-2   Basic RS latch. (a) Cross-connected NOR gates; (b) truth table; (c) logic symbol; (d) NOR gate as viewed by the troubleshooter.

customarily implied (not explicitly shown in the gate's logic symbol). However the troubleshooter knows that each gate is internally connected to V_{CC} and to ground—and he knows that some of the trouble situations that he encounters are caused by "open" or "shorted" connections to V_{CC} or ground inside of the IC package.

From RS to D

Next let us briefly consider the widely used D latch, which is a slight elaboration of the basic RS latch. The logic diagram for a D latch is shown in Figure 12-3. "D" denotes its data-input terminal; (there is only one data-input terminal); "Q" and "$\bar{Q}$" denote its output terminals. These output terminals have the functional relations stated by the truth table.

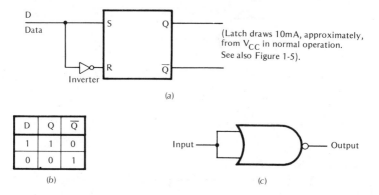

(a)

D	Q	Q̄
1	1	0
0	0	1

(b)

(c)

Note: *The logic diagram in (a) shows that a D latch consists of an RS latch, as depicted in Figure 12-2(c), plus an inverter connected in series with the reset input. Since the inverter is driven from the set input, a D latch has only one input, called the data (D) input. In turn, the truth table shown in (b) is comparatively simple and well suited to practical logic systems. Observe in (c) how a NOR gate is connected to operate as an inverter. Both of its inputs are tied together. If the inverter input goes logic-high, its output goes logic-low; or, if the inverter input goes logic-low, its output goes logic-high.*

Since we are considering TTL logic, this means that if the inverter input is driven +2 volts high, the output will normally go +.09 volt low; or, if the inverter input is driven +.09 volt low, the output will normally go +3.75 volts high, approximately.

Figure 12-3   D latch. (a) Logic diagram; (b) truth table; (c) NOR gate connected to operate as an inverter.

Just what is to be understood by a logic-high designation and by a logic-low designation? In the case of TTL logic, these states are defined as follows:

1. The power-supply voltage has a maximum permissible value of +5.25 volts. Its typical operating value is +5.1 volts.
2. With respect to the gate (or latch) input terminals, any voltage in the range from +2 volts to +5 volts is *normally* responded to as a logic-high level.
3. With respect to the gate (or latch) input terminals, any voltage in the range from zero (ground potential) to +0.8 volt is *normally* responded to as a logic-low level.
4. With respect to the gate (or latch) output terminals, any voltage in the range from +2.4 volts to +5 volts is *normally* responded to as a logic-high level.
5. With respect to the gate (or latch) output terminals, any voltage in the range from zero (ground potential) to +0.4 volt is *normally* responded to as a logic-low level.

We observe, accordingly, that the input state levels are separated by a "bad region" from +0.8 to +2 volts—a range of 1.2 volts. Therefore, if we are tracking down a trouble symptom and we happen to measure +1.6 volts at an input terminal (as we might very well measure in a trouble situation), we will conclude that this input terminal is "in the bad region," and that there is a defect inside of the IC, or outside in the branch circuit (node).

Next we observe that the output state levels are separated by a bad region from +0.4 to +2.4 volts—a range of 2 volts. Therefore, if we are tracking down a trouble symptom and we happen to measure +1 volt at an output terminal (as might very well occur in a trouble situation), we will conclude that this output terminal is "in the bad region," and that there is a defect inside of the IC, or outside in the associated node.

Practical Note

V_{CC} has a normal tolerance, such as from +4.75 to +5.25 volts. This means that if V_{CC} should happen to increase to +6 volts, for example, the IC would be damaged. This also means that V_{CC} may normally operate above the maximum permissible gate (or latch) input voltage. In other words:

1. In the example under consideration, suppose that V_{CC} is operating at +5.25 volts.
2. Since the maximum permissible logic-high potential is +5 volts, V_{CC} is +0.25 volt above the maximum logic-high level. This means that we cannot use V_{CC} as a logic-high test voltage (in this case), because there would be danger of damage to the IC by exceeding its maximum rated input voltage.
3. The troubleshooter can easily avoid this hazard by using a series resistor when making level tests. For example, a gate (or latch) input can safely be connected to a +5.25 volt V_{CC} point *if a 100-ohm series resistor is employed.*

FLOATING INPUT TERMINAL "LOOKS" LOGIC-HIGH

Although it might be supposed that an input terminal to a gate or latch would measure zero volts when the input is "floating," or unconnected to a voltage source or to ground, *this is not so.* As depicted in Figure 12-4, a "floating" input to a D latch typically measures +1.6 volts to ground, and the "floating" terminal "looks"

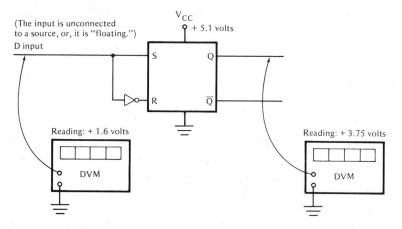

Note: An example of a "floating" input is a cold-solder joint. Another example is a microscopic break in a printed-circuit (PC) conductor, or trace, along the D input line, or node. Observe that as you move step-by-step along the PC conductor with the DVM, its reading will suddenly change from +1.6 to +3.75 volts or to +0.09 volt as you "jump" from one side of the "open" to the other side.

Note: A "bad region" voltage that "looks like" a logic-high level is explained on the basis that a "floating input" is connected neither to a logic-low level nor to a logic-high level—it is an open-circuit condition. It will be explained subsequently that the open-circuit condition cuts off the input transistor in the inverter, just as a logic-high input cuts off the input transistor. In normal operation, the inverter input will be either logic-high or logic-low.

Figure 12-4   Although the D input is in the "bad region," and is neither logic-high nor logic-low, the Q output is logic-high because a "floating" input "looks like a logic-high level" to the latch.

logic-high to the latch. In other words, the Q output of the latch measures +3.75 volts to ground, in this example.

Troubleshooting conclusion: *If you measure +1.6 volts to ground at a D-latch input, look for an open-circuit.* The open circuit will be located prior to the point of measurement (if the open circuit were inside the IC, you would measure either +3.75 volts or +0.09 volt from the preceding driver).

Another troubleshooting conclusion: *If you measure zero volts to ground at a D-latch input, look for a short-circuit to ground.* The short-circuit may be located inside the IC, or it may be located along the associated node, or it may be located inside of the preceding driver IC. For example, you might find a solder splash or "whisker" short-circuiting the PC node conductor to ground.

Still another troubleshooting conclusion: *If you measure V$_{CC}$ at a D-latch input, look for a short-circuit to V$_{CC}$*—logic-high does not normally exceed +3.75 volts. The short-circuit may be located inside of the IC, or it may be located along the associated node, or it may be located inside of the preceding driver IC. As we progress, we will learn how to make tests for pinpointing short-circuit and open-circuit locations.

FAMILIARIZATION EXPERIMENTAL PROJECT

In this experimental troubleshooting project, we construct a D latch from a quad 2-input NOR gate IC package, such as the Radio Shack Type 7402, or equivalent (see Figure 12-5). It is powered by

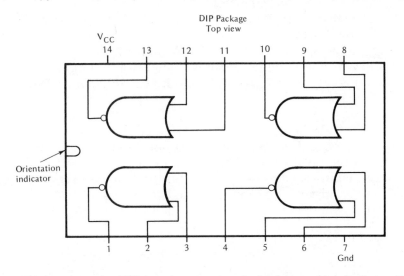

Note: *This is an example of TTL (transistor-transistor logic). It is a multiple logic-gate IC package (quad NOR gate IC package). Since each gate has two inputs, it is termed a quad 2-input NOR gate IC package. The V$_{CC}$ terminal (pin 14) and the gnd terminal (pin 7) are connected inside of the package to each of the NOR gates.*

Note: *Since pins 7 and 14 are connected to each NOR gate inside of the IC package, the V$_{CC}$ current demand depends on the states of all NOR gates, whether they are or are not included in an external circuit. With all input and output pins of the IC package "floating," the V$_{CC}$ current demand is 17 mA, approximately. This is just another way of saying that if only one NOR gate in the package is included in an external circuit, the V$_{CC}$ current demand depends also upon the idling current drain of the three remaining NOR gates.*

Figure 12-5   Package pinout for Type 7402 quad 2-input positive NOR gate.

means of the simple regulated 5.1 volt supply depicted in Figure 12-6. The IC package shown in Figure 12-5 may be mounted and interconnected in any manner that the experimenter prefers. However it is suggested that the quickest and easiest way of constructing the D latch is to use an experimenter's socket, such as the Radio Shack 276-174, or equivalent. In any case, the IC pins are interconnected as shown in Figure 12-7.

Connect the power supply to the D latch, and verify the following voltages:

1. With the D input connected to ground (pin 7), the normal voltage at pin 1 is +0.09V; at pin 2, +3.75V; at pin 3, +3.75V; at pin 5, 0V; at pin 7, 0V. (Logic-high and logic-low values have a ±10 percent tolerance.)
2. With the D input connected through a 100-ohm resistor to V_{CC} (pin 14), the normal voltage at pin 1 is +3.75V; at pin 2, +0.09V; at pin 3, +0.09V; at pin 5, 5.00V; at pin 7, 0V.
3. With the D input "floating," the normal voltage at pin 1 is +4.06V; at pin 2, +0.09V; at pin 3, +0.09V; at pin 5, 1.61V; at pin 7, 0V.

Note that when the D input is "floating," the resulting logic states are the same as if the D input were connected to V_{CC}. On the other hand, some details of circuit action are a bit changed. Thus, the voltage at pin 1 is +4.6V, instead of +3.75V; the voltage at the D input is +1.61V, although no external voltage is applied.

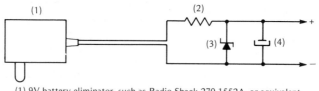

(1) 9V battery eliminator, such as Radio Shack 270-1552A, or equivalent.
(2) 90-ohm ¼ watt resistor.
(3) Zener diode, such as Radio Shack 276-565, or equivalent.
(4) 500μF 16V electrolytic capacitor.

Note: *A 5.1V zener diode operates as a shunt regulator. Observe that the diode is reverse-biased; the positive line from the battery eliminator connects to the series resistor, and thence to the cathode of the zener diode.*

Figure 12-6 Configuration of the 5.1 volt regulated power supply.

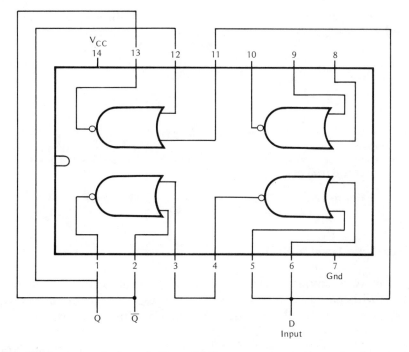

Note: *The power supply shown in Figure 12-6 is connected to the V$_{CC}$ and gnd terminals of the IC. The +5.1V output is connected to pin 14, and the −5.1V output is connected to pin 7. (Pins 8, 9, and 10 are not connected, and this fourth gate is not used in this experiment.)*

Figure 12-7   IC 7402 interconnects for D-latch operation.

DMM With Built-In Level Detector
Operates as a Digital Logic Probe

If a digital multimeter (DMM) with a built-in level-detector function is available, repeat the foregoing measurements using the level director as a digital-logic probe. Observe that all circuit voltages over +0.8V are indicated by an "up" arrow; all circuit voltages under +0.8V are indicated by a "down" arrow; when a circuit voltage is at ground potential, the "down" arrow may be accompanied by a fluctuating minus sign. (See Figure 12-8.)

As a practical note, the troubleshooter should remember that a simple logic probe, such as a level detector, cannot distinguish between a "bad" level and a logic-high level. When it is suspected that a "bad" level may be present, dc voltage measurements should be made.

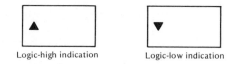

Logic-high indication   Logic-low indication

Note: *A DMM with built-in level indicator may also provide an optional tone indication. For example, when the logic level is low, a "down" arrow is displayed and an audible tone is also produced. On the other hand, when the logic level is high, an "up" arrow is displayed, but no audible tone is produced.*

Figure 12-8   Logic-high and logic-low indications by level detector.

Dual-Tone Level Indicator

You may sometimes want to check logic levels on the basis of tone only, without any visual display. This can easily be accomplished by use of a small feedback amplifier, such as the Radio Shack Mini Speaker Amplifier, No. 277-1008A. To convert the miniature amplifier into a tone generator, a 50-pF capacitor is connected from the blue-lead terminal of the mini speaker to the external-input plug. In turn, when the volume control is advanced, the speaker produces a low-frequency tone.

To operate this feedback amplifier as a dual-tone logic-level indicator, a 100-kilohm resistor is used as a logic probe. The resistor is connected to a lead, which is in turn connected to pin 8 of the integrated circuit inside the amplifier. Logic-level tests are made as follows:

1. Connect the ground side of the external-input plug from the amplifier to the negative power-supply terminal (ground terminal) of the logic circuit under test.
2. Touch the 100-kilohm probe resistor to the negative power-supply terminal (gnd) of the logic circuit under test, and adjust the volume control on the feedback amplifier to a point where the speaker tone slows down to a stop.
3. Now touch the probe resistor to a logic-high point in the circuit—a comparatively high tone will be heard from the speaker.

4. Then touch the probe resistor to a logic-low point in the circuit—a comparatively low tone will be heard from the speaker.
5. If the probe resistor is touched to a circuit point which is at ground potential, the speaker will be silent.
6. If the probe resistor is touched to a "bad-level" point in the circuit, a medium-high tone will be heard from the speaker.

Note that almost any audio amplifier can be experimentally modified to function as a dual-tone logic-level indicator. In other words, a small capacitor can be connected from a point in the output circuit to a point in the input circuit which provides positive feedback and self-oscillation. In turn, when a logic-level voltage is "bled" through a probe resistor into an RC-coupled base circuit of the amplifier, this bias shift results in an output tone change.

Basic LED Level-Indicator Probe

A simple LED level-indicator probe is diagrammed in Figure 12-9. This is a "piggy back" arrangement that is powered from the logic circuit under test. It comprises an LED connected in series with a current-limiting resistor and a Darlington-pair switch. The transistors are cut off when the probe tip is "floating" or when it is connected to a point at ground (or near-ground) potential. On the other hand, when the probe tip is connected to a logic-high point, the transistors are biased into conduction, and the LED glows.

This is basically a go/no-go type of indicator. It shows whether a logic circuit is operating at a correct level, but it cannot provide data concerning why a circuit is malfunctioning. Accordingly, if an LED probe indicates that an incorrect level is present at a particular test point in a logic circuit, the troubleshooter must then proceed to make dc voltage measurements.

Consider a trouble situation in which the Q output of a D latch happens to be short-circuited to ground. An LED probe shows that the Q node remains logic-low, although the D input is being driven logic-high. It is concluded that a fault is present. Next the troubleshooter should measure the dc voltage at the Q node—it will measure zero in this example, instead of +0.09 volt, thereby pointing to a short-circuit to ground, either inside of the IC or along the PC conductor.

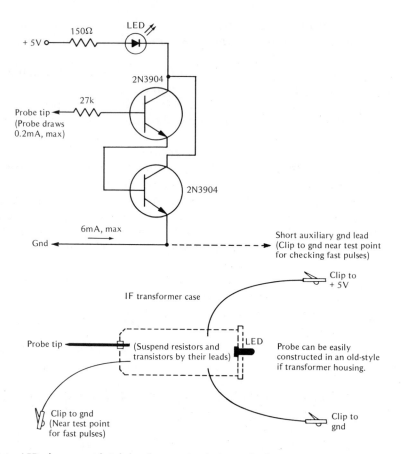

Note: *LED glows most brightly when probe tip is touched to a V_{CC} point (+5.1 volts).*
LED glows brightly when probe tip is touched to a logic-high point (+3.75 volts).
LED glows dimly when probe tip is touched to a "bad-level" point (+1.61 volts).

Figure 12-9   Simplest LED logic-probe arrangement.

CIRCUIT ACTION OF NOR GATE

Before describing other D-latch logic tests, it is helpful to note the circuit action of a NOR gate. It typically comprises six transistors and a diode, as shown in Figure 12-10. Observe that:

1. Input A is being driven logic-high, and input B is being driven logic-low.

2. The base-emitter junction in Q1 is reverse-biased; the base-emitter junction in Q3 is forward-biased.
3. Since the collector-base junction of Q1 is forward-biased, Q2A and Q4 also develop forward-biased collector-base junctions and heavy current flows as shown by the thick lines.
4. The base-emitter junction of Q3 also conducts a heavy current, inasmuch as its emitter terminal is being driven logic-low.
5. Because Q4 is saturated, its collector voltage is very low, with the result that the load goes logic-low.

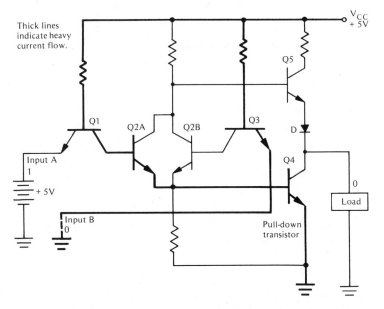

Note: This is an example of a static test; in other words, the test is made with constant dc levels (+5V and ground potential). Although a static test is informative, it is not conclusive. If a gate fails static tests, it is faulty and should be discarded. However, if a gate passes static tests, it may still malfunction when driven by a digital pulse train at rated "clock" frequency.

Figure 12-10   TTL NOR gate; a logic-high input produces a logic-low output.

Next observe the circuit action that occurs when both of the NOR gate inputs are driven logic-low, as depicted in Figure 12-11:

1. The base-emitter junctions of Q1 and Q2 are both forward-biased, and heavy currents flow from their emitters to ground.

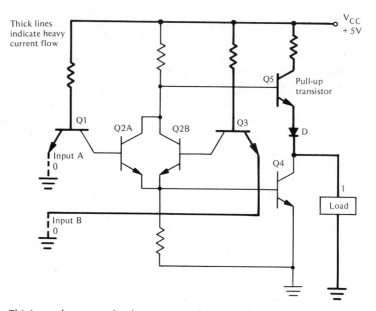

Note: *This is another example of a static test. The gate is defective if it fails the static test, and should be discarded. On the other hand, if the gate passes the static test, it might still fail a dynamic test. As detailed subsequently, a dynamic test can be made with a square-wave generator operating at rated "clock" frequency, and with a digital logic probe as indicator.*

Figure 12-11   TTL NOR gate; logic-LOW inputs produce a logic-HIGH output.

2. In turn, the collector voltages of Q1 and Q3 are very low, with the result that Q2A and Q2B are both cut off.
3. Since there is no current flow through Q2A and Q2B, Q4 cuts off; however the collector voltage on Q2A and Q2B is high, with the result that Q5 is driven into saturation.
4. Because Q5 saturates, it conducts a heavy current through diode D and the load.
5. Heavy current flow through the output load causes it to go logic-high.

Ohmmeter Tests of NOR Gate

Inspection of the NOR-gate circuitry will show that both of the input terminals and the output terminal will normally exhibit diode action with respect to the ground terminal. In other words, if the positive lead of a hi-pwr ohmmeter is connected to the ground terminal of the NOR gate, and the negative lead is connected to the output terminal, a low forward resistance is normally measured. Similarly, low forward-resistance values are normally measured from each of the input terminals to ground. *A high or an infinite measured value indicates that the NOR gate is defective.*

Next, when the ohmmeter test leads are reversed, virtually infinite resistance is normally measured from the output terminal to ground, and from each of the input terminals to ground.

A cross-check can be made with the ohmmeter operated on its lo-pwr ohms function. If the positive lead of the lo-pwr ohmmeter is connected to the ground terminal of the NOR gate, and the negative lead is connected to the output terminal, virtually infinite resistance is normally measured. Similarly, practically infinite resistance values are normally measured from each of the input terminals to ground. Any significant resistance reading is sufficient reason for discarding the NOR gate as defective.

Caution: If a hi-pwr ohmmeter is used that operates with a 9-volt battery on its Rx10,000 range, a good NOR gate will *seem* to fail the reverse-resistance test. This "false alarm" is due to zener action from the comparatively high-test voltage. To avoid this misleading situation, operate the ohmmeter on its Rx1000 range.

D LATCH SOURCE RESISTANCE REQUIREMENT

This simple experiment demonstrates the source resistance range that is required for normal operation of a D latch. The procedure is as follows:

1. With reference to Figure 12-7, connect a dc voltmeter at the Q output.
2. With the D input "floating," note that the Q output is logic-high.
3. Connect the D input through a 2-kilohm potentiometer to ground; this potentiometer functions as the internal resistance of the logic-low source.
4. As the potentiometer resistance is reduced, a point will be noted at which the Q input goes logic-high (voltmeter reading rapidly increases from +0.09V to +3.75V).
5. Disconnect the potentiometer and measure its resistance; this value will be approximately 600 ohms.

Accordingly, we recognize that TTL circuitry is low-impedance circuitry. A driving source must have an internal impedance of less than 600 ohms to ensure normal operation of the driven circuit. In *normal* operation, a TTL circuit has a dynamic internal resistance in the range from 100 to 200 ohms.

Practical example: The logic-high level at the output of the foregoing D latch measures +3.75V. When a 1000-ohm resistor is shunted from the output terminal to ground, the logic-high level is reduced to 3.31V. This is a decrement of 0.44V at a current flow of 3.31 mA. Accordingly, the dynamic internal resistance is equal to 133 ohms.

**Getting the Most Out of Your
Digital-Logic Test Equipment**

Digital troubleshooters sometimes tend to overlook the total capabilities of basic (and inexpensive) test equipment. We will find that even a VOM has unsuspected capabilities in digital troubleshooting procedures, particularly when used with simple supplementary equipment. Faster and easier troubleshooting is based on understanding the total capabilities of your test equipment.

CHAPTER 13

BASIC COUNTER TROUBLESHOOTING

TOGGLE LATCH OPERATION • TOGGLE LATCH TRIGGER ACTION • QUICK CHECK OF LATCH OPERATION • "BOUNCE" TROUBLE SYMPTOM • SHORT-CIRCUIT LOCALIZATION • BASIC 4-BIT BINARY COUNTERY • OPERATION OF CLEAR LINE • NULLING CONSIDERATIONS • SLOW PULSE GENERATOR

TOGGLE LATCH OPERATION

A basic binary counter configuration consists of a series of toggle latches, such as those shown in Figure 13-1. This is an example of *discrete* logic—in other words, the circuit comprises individual transistors, diodes, resistors, and capacitors. Troubleshooters encounter discrete logic on occasion in comparatively simple applications. The arrangement depicted in Figure 13-1 is very instructive for the beginning troubleshooter to construct and to check out as explained below.

The toggle latch has a Q output terminal and a $\overline{Q}$ output terminal. When a voltage source, such as a 6V lantern battery is connected between the V_{CC} and ground terminals, the outputs will exhibit complementary logic states. In other words, if Q is logic-high, then $\overline{Q}$ will be logic-low. Or, if Q is logic-low, then $\overline{Q}$ will be logic-high. DC voltage measurements at the outputs will show that logic-high is normally 5 volts, and logic-low is normally 0.12 volt, approximately. (See Chart 13-1.)

TOGGLE LATCH TRIGGER ACTION

Consider next how this toggle latch may be tested by manual triggering. If a test lead is connected to the trigger input terminal, the latch will change state as follows:

1. Touch the test lead to V_{CC}; this places a positive charge on the 1000-pF capacitors.
2. Next touch the test lead to the ground terminal; the 1000-pF capacitors are then discharged.

193

3. Discharging the 1000-pF capacitors causes the toggle latch to "toggle," or to change its output states—this action is shown by a dc voltmeter connected at the Q output terminal (or the $\overline{Q}$ output terminal).

4. Each time that steps 1 and 2 are repeated, toggling occurs, and the output states are reversed.

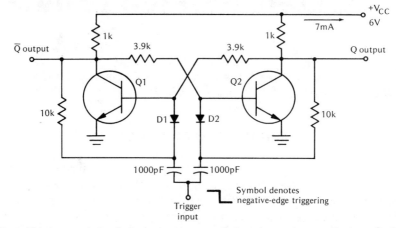

Note: This is a toggle latch; it changes state each time that a trigger pulse is applied. In turn, there is one output pulse for every two trigger pulses. The toggle latch is an elementary binary counter; it counts to 2 and then starts over again. It is a negative edge-triggered latch—the Q output changes from logic-high to logic-low (or from logic-low to logic-high) when the trigger voltage falls from +6V to zero. This negative edge-triggering action is provided by the steering diodes D1 and D2.

Note: A lantern battery such as the Eveready 1209 is suitable for operating the toggle latch. A fresh battery will measure approximately 6.60 volts on open circuit. When a 100-ohm resistor is connected across the battery terminals, the voltage drops typically to 5.85 volts. The current demand is 59 mA, approximately, in this example. Accordingly, the internal resistance of the battery is equal to $0.75/0.059$, or 13 ohms, approximately. If the current demand increases, the internal resistance also increases. As the battery becomes weaker, its internal resistance further increases.

Figure 13-1   Toggle latch configuration.

Chart 13-1

QUICK CHECK OF LATCH OPERATION

A useful quick check of latch operation can be made by improvising a *single-shot pulse generator* from a 0.05-μF fixed capacitor and a test lead:

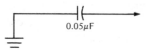

Chart 13-1 *continued*

First suppose that the Q output is logic-high (dc voltmeter reads approximately +5 volts):

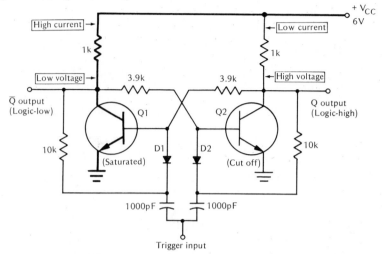

As shown above, one end of the fixed capacitor is connected to ground. The other end is connected to a test lead. Touch the test lead to V$_{CC}$ (this charges the capacitor). Then touch the test lead to the base of Q2. Normally, Q2 will now saturate, and Q1 will cut off. Otherwise, there is a circuit malfunction.

Second, suppose that the Q output is logic-low (dc voltmeter reads approximately +0.12 volt):

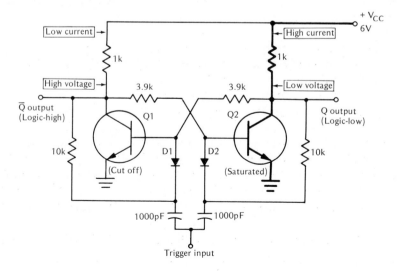

Chart 13-1 *continued*

Touch the test lead to V_{CC} as before, and then touch the test lead to the base of Q1. Normally, Q1 will now saturate, and Q2 will cut off. Otherwise, there is a circuit malfunction.

In case of malfunction, dc voltage measurements can be supplemented by in-circuit resistance measurements with a lo-pwr ohmmeter. Common causes of malfunction are open circuits, short-circuits, and leaky semiconductor junctions. Capacitors may become leaky, and resistors may drift off-value.

When the 1000-pF capacitors are discharged, a *negative edge trigger* pulse is produced. This is just another way of saying that a negative-going trigger pulse is applied to diodes D1 and D2. These diodes operate in combination with the transistor collector-base voltages to momentarily drive the conducting transistor into nonconduction. In turn, the nonconducting transistor is driven into conduction, and the latch locks. The latch remains locked until another trigger pulse arrives, whereupon it again reverses its output states.

"BOUNCE" TROUBLE SYMPTOM

Digital troubleshooters occasionally encounter "bounce" trouble symptoms. This type of malfunction is usually caused by mechanical switching facilities. Consider the experimental arrangement shown in Figure 13-2. This is a toggle latch configuration with the trigger input terminal connected to +V_{CC} through a 100-kilohm resistor. It appears reasonable to suppose that when the test lead is momentarily touched to ground, that the latch will change state.

When this experiment is made, it will be observed that the latch changes its state—*at least once, and often several times in very rapid succession.* This uncertainty results from the fact that "clean" contact is seldom made (see Figure 13-2). Accordingly, the latch often toggles several times before the trigger input is securely grounded. As a result, the final state of the latch output is unpredictable, as can be checked by connecting a dc voltmeter to the Q output (or to the $\overline{Q}$ output).

In the toggling operation first described, "bounce" was less troublesome because the 1000-pF capacitors were initially charged,

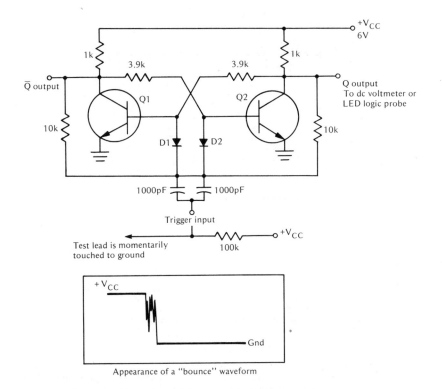

Figure 13-2   Experimental arrangement that demonstrates a "bounce" trouble symptom.

and the charging lead was then disconnected from V_{CC}. Next, when the 1000-pF capacitors were discharged, there was only a small probability that the latch would toggle more than once (however, this possibility does exist, and you may have noted bounce occasionally).

Electronic debouncing circuits are included in many digital systems to eliminate the possibility of false triggering. We will find that the central requirement is to "lock out" the trigger input circuit promptly after arrival of the trigger wavefront. Thereby, subsequent contact "chatter" cannot cause false triggering.

SHORT-CIRCUIT LOCALIZATION

Digital troubleshooters are often concerned with locating a short-circuit. For example, with reference to Figure 13-3, if there is a short-circuit along a PC conductor to the Q output (collector of Q2),

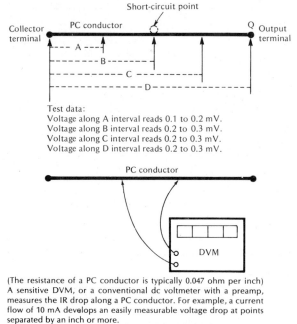

Test data:
Voltage along A interval reads 0.1 to 0.2 mV.
Voltage along B interval reads 0.2 to 0.3 mV.
Voltage along C interval reads 0.2 to 0.3 mV.
Voltage along D interval reads 0.2 to 0.3 mV.

(The resistance of a PC conductor is typically 0.047 ohm per inch)
A sensitive DVM, or a conventional dc voltmeter with a preamp,
measures the IR drop along a PC conductor. For example, a current
flow of 10 mA develops an easily measurable voltage drop at points
separated by an inch or more.
(If your DVM has inadequate sensitivity,
refer to Figs. 10-1, 10-2, and 10-3.)

When a short-circuit occurs, the effective value of the circuit load is generally reduced considerably. This is just another way of saying that the short-circuit current will be much greater than the normal current.

Figure 13-3   Example of short-circuit localization.

the Q output will be "stuck low" and cannot be driven high. The voltage from collector to ground, which normally measures +0.12V in the logic-low state, now measures zero. The troubleshooter then knows that a short-circuit is present, although he does not know where the short-circuit is located.

In this example, a short-circuit in the collector-output circuit results in a current flow of approximately 10 mA. This current will flow through the transistor to ground, if the transistor is shorted. Or, if the transistor is normal, this 10 mA of short-circuit current will flow through the Q-output PC conductor. Accordingly, the trouble-shooter needs to know whether this current is flowing through the PC conductor, or whether the PC-conductor current is zero.

In this situation, the question can be answered (in most cases) by means of a voltage measurement with a sensitive DVM. The troubleshooter proceeds as follows:

1. The DVM test leads are applied at the collector terminal of Q1 (Figure 13-3) and at the Q output terminal on the PC board.
2. A reading of several tenths of a millivolt indicates that the short-circuit current is flowing through the PC conductor.
3. A reading of zero indicates that the short circuit is in the transistor, or that the short circuit is not far from the collector terminal of the transistor.

Case history: A toggle latch with the configuration shown in Figure 13-2 developed a "stuck low" trouble symptom at the Q output. The Q terminal voltage to ground measured zero. Next, the troubleshooter operated the DVM on its millivolt range, and measured the voltage drop along the PC conductor from the collector terminal of Q2 to the Q output terminal. The DVM indicated that a voltage drop of 0.2 or 0.3 mV was present (the readout fluctuated slightly).

Next, the troubleshooter started probing back from the Q output terminal toward the collector terminal of the transistor. There was no change in DVM reading until the probe was moved within 1¾ inches of the collector terminal—then the reading decreased to 0.1 mV, and finally to zero. Therefore, the troubleshooter concluded that the short-circuit was located more than 1 inch, but less than 2 inches, down the PC conductor from the collector terminal.

A close inspection showed that a short-circuit within the localized interval had occurred as the result of solder "spread" from an adjoining pad—when the pad was soldered to a replacement component, a short-circuit had accidentally been made to the Q output PC conductor—the solder "spread" was covered by a flow of varnish and was barely visible. (See Figure 13-3.)

BASIC 4-BIT BINARY COUNTER

Consider next the 4-bit binary counter configured from toggle latches as depicted in Figure 13-4. This is an *asynchronous* counter, inasmuch as it is not clocked. Asynchronous counters are also called *serial* or *ripple* counters. The term "serial" means that data (trigger pulses) are applied to Latch 1, after which Latch 2 is triggered from Latch 1, and so on. The term "ripple" means that the carry bit from Latch 1 is passed into Latch 2, from which it may be passed into Latch 3, and so on.

Suppose, for example, that seven trigger pulses have been applied to Latch 1. In turn, the count at the Q outputs is 0111. Now, if

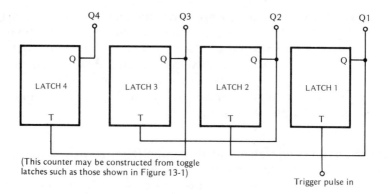

(This counter may be constructed from toggle latches such as those shown in Figure 13-1)

Trigger pulse in

Note: The toggle latches are shown cascaded from right to left, so that readouts will correspond to standard binary-number notation. For example, the count progresses as follows:

1. *No Input Pulses: Readout = 0000*
2. *One Input Pulse: Readout = 0001*
3. *Two Input Pulses: Readout = 0010*
4. *Three Input Pulses: Readout = 0011*
5. *Four Input Pulses: Readout = 0100*
6. *Five Input Pulses: Readout = 0101*
7. *Six Input Pulses: Readout = 0110*
8. *Seven Input Pulses: Readout = 0111*
9. *Eight Input Pulses: Readout = 1000*
10. *Nine Input Pulses: Readout = 1001*
11. *Ten Input Pulses: Readout = 1010*
12. *Eleven Input Pulses: Readout = 1011*
13. *Twelve Input Pulses: Readout = 1100*
14. *Thirteen Input Pulses: Readout = 1101*
15. *Fourteen Input Pulses: Readout = 1110*
16. *Fifteen Input Pulses: Readout = 1111*
17. *Sixteen Input Pulses: Readout = 0000*

This is an "up" counter, because the readout increases with each trigger.

(Counter "overflows" and returns to zero.)

Figure 13-4   Binary counter configured from toggle latches.

one more trigger pulse is applied to Latch 1, Q1 goes to 0, and a carry bit is passed into Latch 2. In turn, Latch 2 goes to Q2 = 0, and a carry bit is passed into Latch 3. Then Latch 3 goes to Q3 = 0, and a carry bit is passed into Latch 4. Thereupon Latch 4 goes to Q4 = 1, and the final readout becomes 1000.

Probe Readout

If you construct and experiment with this counter configuration, you may wish to construct four LED logic probes, as was described in

Figure 12-9. When a probe is connected at each of the Q outputs in Figure 13-4, a visible counter readout is obtained. (Note that all of the probes and latches are powered from the same 6V lantern battery.)

When the V_{CC} voltage is first applied, the counter is not likely to read 0000. The reason for the unpredictable starting count is in tolerances on transistors, diodes, resistors, and capacitors. Therefore the first requirement is to clear the counter, and thereby make it read 0000. At this time we will clear the counter by manual triggering; later, we will employ a "clear" or "reset" line to automatically start the counter with a 0000 readout.

To clear (reset) the counter manually, proceed as follows:

1. Connect a test lead to the "Trigger Pulse In" terminal of Latch 1 in Figure 13-4.
2. Touch the other end of the test lead alternately to $+V_{CC}$ and then to gnd.
3. Each time that the test lead is touched to gnd, the count will advance by 1.
4. Finally, the counter will read 1111; then the next time that the test lead is touched to $+V_{CC}$ and then to gnd, the counter will read 0000.
5. The counter is now cleared, or reset, and it will count up from zero when subsequent trigger pulses are applied.

To review briefly, "clear" means to return the counter readout to 0000. "Reset" means that the counter readout is returned to some pre-determined value such as 0000, or to some other predetermined value. Thus, "reset" is a more general term than "clear."

OPERATION OF CLEAR LINE

To reset the toggle latch (return the Q output to a logic-low state), the base of Q2 is momentarily driven positive. A diode and a 1-kilohm resistor are utilized, as depicted in Figure 13-5. The diode serves to isolate the base from the Reset line, unless a positive voltage is applied to the Reset terminal. The 1-kilohm resistor is a current limiter; it prevents burnout of the diode D3 and transistor Q2.

Next, this basic principle is employed in connection of a Clear line to the counter, as shown in Figure 13-6. Note that each of the toggle latches has been provided with an isolating diode (D3), which in turn is connected to the Reset terminal (R). All four Reset terminals are tied together and connected to a 1-kilohm resistor, which is in turn connected to the Clear terminal.

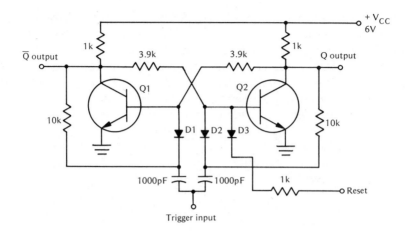

Note: To reset (clear) the latch, and make Q = 0, the Reset terminal is driven positive by application of +6V. If the Q output is logic-high, the Reset voltage will drive it logic-low. If the Q output is already logic-low, it will remain logic-low. After the reset voltage is removed, the latch may be operated from its trigger input.

Figure 13-5   Toggle latch is reset (cleared) by driving the base of Q2 positive.

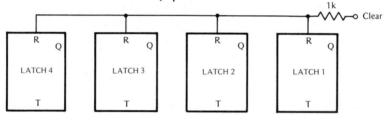

Note: Each R (Reset) terminal connects to D3 in the associated latch, as depicted in Figure 13-5. In other words, only one 1k current-limiting resistor is used in the clear circuit. To clear the counter (reset each Q output to 0), the Clear terminal is momentarily connected to V_{CC}. Observe that the remainder of the counter wiring is shown in Figure 13-4.

Troubleshooting hint: When a counter malfunctions, measure V_{CC} first. If V_{CC} becomes subnormal, the counter will continue to operate but the count will be incorrect, depending upon device and component tolerances.

Figure 13-6   Connection of a Clear line to the 4-bit binary counter.

If you construct this arrangement for experimenting, the Clear line can be manually driven. Connect a test lead to the Clear terminal. To clear the counter, touch the test lead to a V_{CC} terminal. If you have connected LED logic probes at the Q outputs, the LEDs will all go dark when the counter is cleared. Then, as binary numbers are entered (trigger input terminal of Latch 1 is pulsed), the LEDs will successively glow to indicate the readout.

NULLING CONSIDERATIONS

Troubleshooters do not expect that a DVM will be precisely nulled on every dc voltage range. Thus, a professional DVM may normally indicate 0.1 mV (instead of 000.0) when its test leads are short-circuited together. Again, a service-type DVM may normally indicate 1 or 2 mV (instead of .000) when its test leads are short-circuited together. In other words, a DVM may normally display a small offset voltage when its test leads are short-circuited together. The DVM can be precisely zeroed, in any case, when used with a preamp such as that described previously. Thus the nulling procedure serves to cancel out both the op-amp offset voltage and the DVM offset voltage. (See Chapter 10.)

Note also that precise zeroing is not necessary in many practical situations. For example, in Figure 13-3, the troubleshooter is not basically concerned with the exact voltage drop along the PC conductor. *The practical consideration is to locate the point along the conductor where the voltage indication no longer increases, but remains constant, as the test prod is moved farther down the conductor.*

SLOW PULSE GENERATOR

Preliminary troubleshooting procedures may employ manual pulsing of a counter, as previously described, or slow automatic pulsing may be used, as shown in Figure 13-7. This multivibrator arrangement is a square-wave generator with a repetition rate of approximately 1 pulse per second. When used with the counter depicted in Figure 13-4, negative-edge triggering occurs, and Latch 1 changes state approximately once every second. LED logic probes will provide convenient counter readout.

Note that in normal operation, the slow pulse generator does not produce false triggering ("bounce"). Thus it provides a more reliable input trigger waveform than does manual pulsing. If the trouble-shooter wishes to obtain a higher output frequency, the coupling capacitors in the generator can be reduced in value. Thus 10 μF capacitors will provide approximately 2 pulses per second.

The experimenter may have a square-wave generator available which will serve as a slow pulse generator. Most square-wave generators have a sufficiently rapid fall time to trigger the counter properly. However, square-wave generators that have comparatively slow fall time will trigger the counter erratically, or not at all. This topic is explained in greater detail in the next chapter.

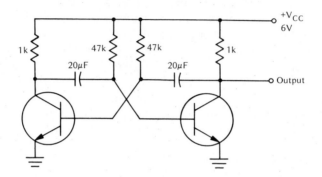

Note: If a "pause" control is desired, connect an SPST pushbutton switch in series with the output lead. Then, the counter under test will "hold" as long as the switch is open. The pulse generator may be powered from the same lantern battery that powers the counter.

Figure 13-7   Slow pulse generator configuration.

CHAPTER 14

ADDITIONAL COUNTER TROUBLESHOOTING TECHNIQUES

DOWN COUNTER OPERATION • CLOCK OPERATION OF BINARY COUNTER • MALFUNCTION RESULTING FROM MARGINAL V$_{CC}$ SUPPLY VOLTAGE • MALFUNCTION CAUSED BY MARGINAL TRIGGER AMPLITUDE • MALFUNCTION CAUSED BY FAST CLOCK • MALFUNCTION CAUSED BY SLOW FALL TIME • MALFUNCTION CAUSED BY INADEQUATE FILTERING • GENERAL GUIDELINES • BYPASSING AND DECOUPLING ARRANGEMENTS • QUICK TEST FOR CAPACITANCE VALUES OF LARGE FILTER CAPACITORS

DOWN COUNTER OPERATION

A down counter is very similar to an up counter, except that the down-counter readout decreases from 1111 to 0000, whereas the up-counter readout increases from 0000 to 1111. For example, when the up-counter arrangement in Figure 13-4 is configured as shown in Figure 14-1, the readout will decrease from 1111 to 0000 on successive trigger pulses. Note that the $\overline{Q}$ output drives the trigger input of the toggle latches in a down-counter configuration.

This down-counter arrangement may be experimentally triggered by connecting a test lead to the trigger-pulse input terminal, and touching the test lead alternately to V$_{CC}$ and then to ground, or, it may be driven by the slow pulse generator depicted in Figure 13-7. This is an *asynchronous* binary counter, inasmuch as it is not clocked. It is also termed a *serial* or a *ripple* down counter. (See Chart 14-1.)

MALFUNCTION RESULTING FROM MARGINAL V$_{CC}$ SUPPLY VOLTAGE

Counter malfunction will result from marginal V$_{CC}$ supply voltage.

Example: The arrangement shown in Figure 14-1 was driven from a slow pulse generator, and the V$_{CC}$ voltage was reduced until the

205

Chart 14-1

CLOCK OPERATION OF BINARY COUNTER

A digital clock is a pulse generator (commonly a square-wave generator) which operates as a synchronizing or pacing signal source for a digital system. A clock oscillator may operate with great rapidity, as in a high-speed computer, or it may operate very slowly, as in a scanner-monitor radio. The troubleshooter will encounter:

1. *Asynchronous logic* in which the operating speed depends only on the speed of signal propagation through the digital network (as in Figure 14-1).
2. *Synchronous logic* in which the logical operations occur in synchronism with clock pulses.

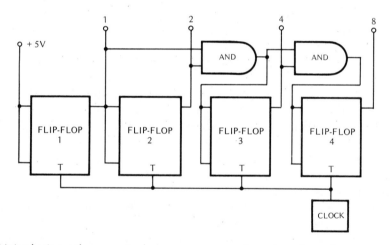

This is a basic synchronous ripple-carry counter configuration, shown for comparison purposes. Its detailed operation is explained subsequently.

However, the troubleshooter will also encounter "hybrid" systems that employ both asynchronous logic and synchronous logic. For example, the counter depicted in Figure 14-1 may be driven by trigger pulses that occur at random intervals, or, it may be driven by clock pulses. In the latter case, only Latch 1 is a synchronized-logic device—Latches 2, 3, and 4 operate as asynchronous logic devices. Their speed of operation depends entirely upon the propagation time through the latch circuits. When going from a 0111 count to a 1000 count, the carry

Chart 14-1 *continued*

bit must "ripple through" Latch 2, and then Latch 3, and finally Latch 4, before the final readout is displayed.

By way of comparison, in following chapters we will work with various clocked bistable multivibrators, called flip-flops (instead of latches). When flip-flops are configured as counters, each flip-flop is clocked, as seen in the foregoing diagram. This type of counter is generally termed a "synchronous counter."

However, in one significant sense of the term, this "synchronous counter with ripple carry" may be regarded as a "hybrid" system inasmuch as the carry bits propagate through the series-connected AND gates, with the result that this delay involved in the ripple-carry action reduces the maximum speed of operation, compared with the operating speed of a true synchronous counter.

---

counter displayed a trouble symptom: "stuck at" 1111. (The slow pulse generator ceased operation.) At this point the V_{CC} voltage measured 2.8V. Note that various other trouble symptoms may occur in this situation, depending upon device and component tolerances.

MALFUNCTION CAUSED BY MARGINAL TRIGGER AMPLITUDE

Counter malfunction will also result from marginal trigger amplitude.

Example: The arrangement in Figure 14-1 was triggered from a bias box; a test lead connected to the trigger pulse input terminal was alternately touched to the positive and to the negative terminals of the bias box. When the bias-box voltage was reduced below 4V, trigger action became erratic; at 3.5V, no trigger action occurred.

Caution: When making peak-voltage measurements of digital pulses with a peak-reading probe (Figure 14-2), remember that the probe "sees" a dc pulse as if it had no dc component. Therefore, the pulse voltage is equal to the *sum of the voltages measured with a positive-peak probe and a negative-peak probe.* Or, the pulse voltage is equal to the voltage measured with a peak-to-peak probe (Figure 14-3).

Note also that germanium diodes are used in the probes depicted in Figure 14-2. Accordingly, the indicated peak-voltage value will be approximately 0.25 volt less than the actual value; the indicated peak-

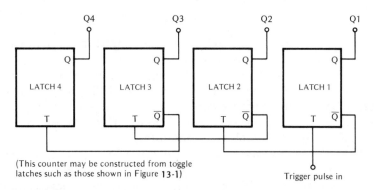

(This counter may be constructed from toggle
latches such as those shown in Figure 13-1)

Trigger pulse in

Note: *The toggle latches are shown cascaded from right to left, so that readouts will correspond to standard binary number notation. For example, the count decrements as follows:*

1. *No Input Pulses: Readout = 1111*
2. *One Input Pulse: Readout = 1110*
3. *Two Input Pulses: Readout = 1101*
4. *Three Input Pulses: Readout = 1100*
5. *Four Input Pulses: Readout = 1011*
6. *Five Input Pulses: Readout = 1010*
7. *Six Input Pulses: Readout = 1001*
8. *Seven Input Pulses: Readout = 1000*
9. *Eight Input Pulses: Readout = 0111*
10. *Nine Input Pulses: Readout = 0110*
11. *Ten Input Pulses: Readout = 0101*
12. *Eleven Input Pulses: Readout = 0100*
13. *Twelve Input Pulses: Readout = 0011*
14. *Thirteen Input Pulses: Readout = 0010*
15. *Fourteen Input Pulses: Readout = 0001*
16. *Fifteen Input Pulses: Readout = 0000*
17. *Sixteen Input Pulses: Readout = 1111*

This is a "down" counter, because the readout decreases with each trigger

Figure 14-1   Down binary counter configured from toggle latches.

to-peak voltage will be approximately 0.5 volt less than the actual value.

Rectifier probes are not useful for measuring peak voltages when pulses have a slow repetition rate. For example, if an attempt is made to measure the peak voltage output of the slow pulse generator described above, using a peak-reading or a peak-to-peak reading probe, the voltmeter indication will fluctuate excessively.

A practical method of measuring the peak voltage of a slow pulse waveform is to use a DMM with a peak-hold control. Alternatively, an oscilloscope can be utilized. A conventional DMM can also be used with an auxiliary peak-hold unit, as shown in Figure 14-4, to measure the peak voltage of a slow pulse waveform.

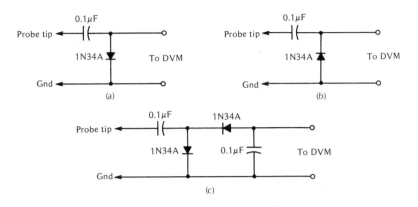

Note: An ac pulse has an average value of zero (it has no dc component). Since all of these probes employ series input capacitors, they do not respond to any dc component that might be present in a pulse waveform (any dc component will be rejected). Most digital-pulse waveforms are dc pulse waveforms, and contain dc component voltages. The total amplitude of a dc pulse is indicated by a peak-to-peak probe. On the other hand, only the positive-peak excursion of the equivalent ac pulse is indicated by a positive-peak probe. Similarly, only the negative-peak excursion of the equivalent ac pulse is indicated by a negative-peak probe. Therefore, if the amplitude of a digital pulse is to be measured, a peak-to-peak probe should be used. Otherwise, if peak probes are used, the indication of the positive-peak probe must be added to the indication of the negative-peak probe.

Note: The indicated peak voltage will be approximately 0.25 volt less than the actual peak voltage, due to the barrier potential of the germanium diode. The indicated peak-to-peak voltage will be approximately 0.5 volt less than the actual peak-to-peak voltage, due to the barrier potential of the germanium diodes. Therefore, 0.25 volt should be added to the peak voltage indicated by the DVM, and 0.5 volt should be added to the peak-to-peak voltage indicated by the DVM.

Figure 14-2   Basic probes used with DVM. (a) Positive-peak probe; (b) negative-peak probe; (c) peak-to-peak probe.

MALFUNCTION CAUSED BY FAST CLOCK

A slow pulse generator is often useful in preliminary trouble-shooting procedures because LED probes can be applied at key test points and sequential responses noted. In other words, the slow pulse indications are individually observable—if the pulse repetition rate is speeded up to the point that persistence of vision occurs, an LED probe will appear to glow continuously, and sequential responses are no longer observable.

Note that any digital system has a maximum trigger-input frequency (maximum clock frequency). *If the maximum permissible*

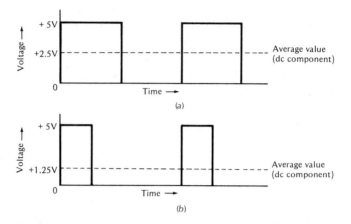

(a)

(b)

Note: *The waveform in (a) will measure 5 volts with a peak-to-peak probe, 2.5 volts with a positive-peak probe, and 2.5 volts with a negative-peak probe. The waveform in (b) will measure 5 volts with a peak-to-peak probe, 3.75 volts with a positive-peak probe, and 1.25 volts with a negative-peak probe. Both of the waveforms are classified as dc-pulse waveforms, inasmuch as their complete excursion has only one polarity.*

Figure 14-3 Examples of digital pulses. (*a*) Square wave with dc component equal to half of total amplitude; (*b*) pulse wave with dc component equal to one-quarter of total amplitude.

clock frequency is exceeded for any reason, the system will malfunction.

Example: With respect to the down binary counter configuration shown in Figure 14-1, the maximum trigger-pulse repetition rate can easily be checked by driving the counter from a square-wave generator. The generator output level is set to approximately 5 volts peak-to-peak. Observe the following points:

1. When the clock is slowed down (generator set to a low repetition rate), LED probes connected at the Q outputs of the counter will flash on and off sequentially, and the action can easily be followed.
2. When the clock is speeded up, the probes appear to glow continuously, and the action cannot be followed.
3. When the clock is further speeded up to approximately 150 kHz (in this example), all of the Q outputs normally "go low." In other words, the apparent malfunction results from a "fast clock."

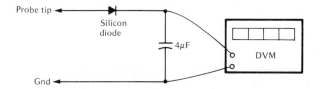

Note: Heavy surge current flow often occurs when the peak-hold unit is initially applied in a circuit. Accordingly, a high-current silicon diode such as the Radio Shack 276-1114 type is used. The fixed capacitor should have very high insulation resistance, such as the GE Pyranol type.

The indicated peak voltage will be approximately 0.6 volt less than the actual value, due to the barrier potential of the silicon diode. Therefore, 0.6 volt should be added to the voltage indicated by the DVM.

This peak-hold unit measures the total amplitude of a dc pulse waveform, because it does not reject the dc component. The peak-hold unit is basically different from a peak-reading probe inasmuch as the peak-hold unit does not employ a series input capacitor.

Figure 14-4   Positive peak-hold unit for DVM.

When all of the Q outputs in the counter go low, all of the $\overline{Q}$ outputs go high. When the clock is operated at 50 kHz, for example, the LED probes appear to glow continuously, due to persistence of vision. *If the Q output voltage is checked with a dc voltmeter, it will measure "half high," or one-half the normal logic-high voltage.* This half-high indication results from the fact that the square-wave voltage is "on" for half the time, and is "off" for half the time. In turn, the dc voltmeter indicates the average value of the dc square waveform.

MALFUNCTION CAUSED BY SLOW FALL TIME

As noted previously, the binary counter arrangement depicted in Figure 14-1 is a form of edge-triggered device (not a level-triggered device). As a result, it will fail to trigger if the pulse edge has too slow a rate of charge (roc). In this example the trailing or negative edge of the trigger input pulse is of concern. If its transition from logic-high to logic-low is too slow, Latch 1 will fail to trigger.

Example: Connect the output from an audio oscillator in series with a 2.5-volt dc source, as shown in Figure 14-5. Apply this "$\sin^2$" waveform to the trigger input pulse terminal of the counter. Set the audio oscillator to an output level of 5 volts peak-to-peak. Then vary the frequency of the $\sin^2$ trigger waveform from zero to 15 kHz.

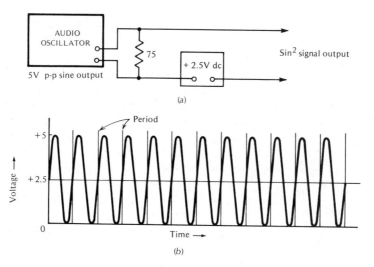

(a)

(b)

Note: *The toggle latch used in this experiment is triggered by +5 volt dc pulses. Accordingly, the +5 volt peak-to-peak sine-wave output from the audio oscillator is connected in series with a +2.5 volt dc source (a). The result is a sine-wave test signal with a +2.5 volt dc component. Thus, the waveform is all positive, with an excursion from zero to +5 volts.*

Technically, the waveform in (b) is termed a sin² waveform. This terminology follows from the mathematical squaring of a sine wave—the result is another sine wave with a dc component as shown in the diagram.

Figure 14-5 Provision of a trigger pulse with slow fall time. (a) Connections; (b) waveform.

Observe that the Q outputs of the counter do not change state in response to the applied trigger waveform—the fall time of the sin² waveform is too slow to actuate the toggle latch used in this experiment.

Next consider the use of a test arrangement that provides a trigger input pulse with an adjustable fall time, so that the requirement of the toggle latch can be measured. A simple integrating circuit is inserted between the output of the slow pulse generator and the trigger pulse input of the counter. We observe that when a 1000-pF capacitor is used with a 1-kilohm resistor in the integrating circuit, the counter triggers normally. On the other hand, if a 2000-pF capacitor is used with the 1-kilohm resistor, the counter does not trigger. (See Figure 14-6.)

When working with integrating circuits, the troubleshooter finds it helpful to consider their time constants. Note the following facts:

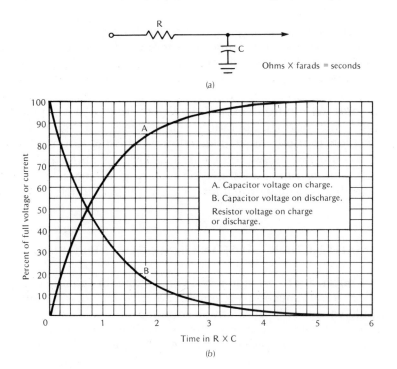

Ohms X farads = seconds

(a)

A. Capacitor voltage on charge.
B. Capacitor voltage on discharge.
Resistor voltage on charge
or discharge.

Percent of full voltage or current

Time in R X C

(b)

Note: When driven by a square-wave input, an RC integrator provides an exponential-wave output. The fall time of the output waveform is usually expressed in time-constant units. The output waveform falls to 37 percent of full voltage in 1 time-constant. An integrating circuit has a time-constant of RC seconds; for example, 1 kilohm and 1000 pF have a time-constant of 1 microsecond. The pulse width of an exponential pulse is equal to its duration at half amplitude. This duration is equal to 0.7 of the time constant. For example, the pulse width of the output waveform from a 1-kilohm and 1000-pF integrating circuit is equal to 0.7 microsecond.

Figure 14-6   RC integrating circuit for controlling fall time. (a) Circuit; (b) universal time-constant chart.

1. The time constant of an integrating circuit is equal to the product of resistance times capacitance; the answer is in seconds.
2. In one time constant, the output from an integrating circuit will fall from maximum voltage to 37 percent of maximum.
3. A 1000-pF capacitor and a 1-kilohm resistor have a time constant of 1 microsecond. A 2000-pF capacitor and a 1-kilohm resistor have a time constant of 2 microseconds.

4. Accordingly, the toggle latch used in this experiment requires a "time-constant fall time" shorter than 2 microseconds. (The latch toggles normally from a time-constant fall time of 1 microsecond.)

Kludge note: When troubleshooting digital circuitry, it is sometimes observed that normal operation will resume if a "jury-rigged" modification is used. As an illustration, a missing count may be restored if the V_{CC} supply voltage is "cranked up" above its normal value. A kludge repair of this type is very poor practice. In this situation, there is a high probability of early call-back and loss of good will.

MALFUNCTION CAUSED BY INADEQUATE FILTERING

It is necessary to provide rated V_{CC} voltage for reliable counter operation. It is also necessary to employ adequate filtering; otherwise, the ripple voltage will cause erratic counting.

Example: With reference to the counter arrangement shown in Figure 14-1, the rated V_{CC} supply voltage is +6 volts. When powered from a lantern battery, counter action proceeds normally. On the other hand, when the counter is powered from a half-wave power supply, sufficient filtering must be utilized to reduce the ripple voltage to less than 0.3 volt rms. *If the ripple voltage exceeds this value, counter action will become erratic, or will stop.*

The ripple voltage in the power-supply output can be measured on the ac function of a DVM. It is impractical to use a VOM in this application because of its comparatively low sensitivity, and also because a very large blocking capacitor must be used in series with the meter test lead.

Internal Resistance of Power Supply

A lantern battery is not an ideal constant-voltage source—it has more or less internal resistance. Similarly, conventional power supplies have more or less internal impedance. The result is that the transient current drains imposed on a battery develop an effective ripple voltage which is synchronized with the counting action. This ripple voltage is less troublesome than a 60-Hz or a 120-Hz ripple voltage.

As an illustration of the difficulty encountered in reducing this synchronized ripple voltage resulting from counter action, the voltage from a 6V lantern battery was experimentally filtered by connecting an 85,000 μF electrolytic capacitor across the battery terminals. This filtering reduced the ripple voltage from 0.3 volt p-p to 0.1 volt p-p, approximately.

GENERAL GUIDELINES

As general guidelines in power-supply troubleshooting, the technician expects to measure:

1. Output voltage under no-load conditions, not to exceed 5.25 volts (standard TTL arrangements).
2. Ripple voltage under maximum current demand, not to exceed 5 percent.
3. Regulation under worst-case conditions, not to exceed 5 percent.

Ripple is defined as the ac component from a dc power supply arising from sources within the power supply. Ordinarily, percent ripple is defined as the ratio of the rms value of the ripple voltage to the absolute value of the total voltage, expressed in percent. Since the ripple waveform is nonsinusoidal, its voltage can be accurately measured only with a true rms voltmeter.

Example: Under maximum current demand, the output voltage from a regulated power supply is 5.1 volts. The ripple voltage, measured on the ac function of a true rms voltmeter is 0.25 volt rms. In turn, the percent ripple is equal 4.9 percent.

Regulation is defined basically as the ratio of the difference between the no-load output voltage and the full-load output voltage to the full-load voltage, expressed in percent.

Example: Under no-load conditions, the output voltage from a regulated power supply is 5.20 volts. Under full-load conditions, the output voltage is 5.09 volts. This is a difference of 0.11 volt. In turn, the percentage regulation is approximately 2.2 percent.

Troubleshooting note: The foregoing example of regulation measurement does not take into account the effect of ac line-voltage variation. Since this is a basic factor in digital system operation, it is essential that the troubleshooter check the power-supply regulation under worst-case conditions of ac line-voltage variation.

Example: With respect to the foregoing example of regulation measurement, the 2.2 percent value corresponds to constant ac line voltage. Now consider the situation in which the line voltage may vary within the limits of 110 and 120 volts. Percentage regulation should be measured with the ac line voltage adjusted to 120 volts at no load, and with the ac line voltage adjusted to 110 volts at full load. The regulation value should not exceed 5 percent.

Experiment: With reference to Figure 12-6, note the following measurements: Line voltage = 110 volts rms; open-circuit output voltage = 5.08 volts; full-load (220 ohms) voltage = 5.00 volts. Line voltage = 120 volts rms; open-circuit output voltage = 5.09 volts; full-load (220 ohms) voltage = 5.04 volts.

We conclude from this experiment that the power-supply regulation under worst-case conditions (in this example) is 2 percent, approximately. Note that the full-load current in this example is approximately 23 mA. It follows that *the internal resistance of this regulated power supply is equal to* $^{0.05}\!/_{0.023}$, *or approximately 2.17 ohms.*

BYPASSING AND DECOUPLING ARRANGEMENTS

Troubleshooters know that digital power supplies are supplemented in most systems by various bypassing and decoupling arrangements. For example, driving/receiving devices are usually decoupled in addition to the normal decoupling provisions—look for 0.1 μF rf capacitors at the V_{CC} and ground pins of driving/receiving devices. Look also for 0.01 to 0.1 μF decoupling rf capacitors in each group of 5 to 10 IC packages. Open capacitors can make the power supply "look bad" because each bypass or decoupling capacitor contributes to ripple reduction and regulation improvement. Note also that you will usually find rf bypass capacitors at the power-transformer primary terminals.

Surge Protection

Surge protection is often provided in addition to bypassing at the power-transformer primary terminals, as depicted in Figure 14-7. The rf bypass capacitor is supplemented by an rfi filter, in this example. A varistor is a two-electrode semiconductor device with a voltage-independent nonlinear resistance characteristic that decreases rapidly in value as the applied voltage increases. Thus, a varistor tends to divert high transient voltages to ground.

Troubleshooters should keep in mind that defective bypassing, rfi filtering, and surge-protecting devices or components can make a power-supply filter "look bad," inasmuch as these ac input arrangements contribute to ripple reduction and regulation improvement.

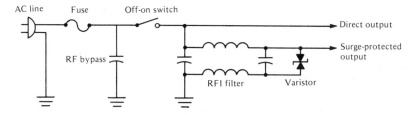

Note: The rf bypass capacitor has a value of 0.22 μF; it reduces the amplitude of rf energy and of sharp transient spikes. This is the only bypassing provided for the direct-output ac line. A branch line is provided with a standard rfi line filter which is terminated by a varistor, such as the GE MOV Series V130LA10A. In turn, the surge-protected output ac line is comparatively free from rf interference and transient surges.

Figure 14-7   AC power input circuit for a small computer.

Of course, a surge protector cannot assist in compensating for sudden dips in ac line voltage—the chief approach to minimizing trouble symptoms resulting from dips is to employ comparatively large values of filter output capacitance.

QUICK TEST FOR CAPACITANCE VALUES
OF LARGE FILTER CAPACITORS

Digital troubleshooters occasionally have the need for a quick check of capacitance values for large filter capacitors. For this purpose a 100-ohm resistor and a 6-volt battery serve the purpose satisfactorily; if a stopwatch is available, the method is highly accurate. If a test jig is devised, a pair of SPST switches can be included, as shown in Figure 14-8. Otherwise, the troubleshooter may simply touch a test lead to the required point.

With reference to Figure 14-6, this capacitance measurement is based on the time constant of the 100-ohm resistor and the filter capacitor. In other words, the capacitor will charge up to 63 percent of the battery voltage in one time constant, or the capacitor will discharge to 37 percent of the battery voltage in one time constant.
Example: A filter capacitor rated for 85,000 μF is checked in the test circuit depicted in Figure 14-8. Sw-2 is first closed to ensure that the capacitor is completely discharged. Then Sw-2 is opened. Sw-1 is then closed, and the time required for the capacitor to charge up to 3.78 volts is noted. In this example, the normal time interval is 8.5 seconds.

A shorter charging time interval indicates that the capacitor does not have its full rated value of capacitance. A longer charging time indicates that the capacitor has more than its rated value of capacitance.

A useful cross-check can be made by opening Sw-1 and then closing Sw-2; the time required for the capacitor to discharge down to 2.22 volts is noted. In this example, the normal time interval is 8.5 seconds.

A *leakage test* is made by charging the capacitor to 6 volts, and then letting it stand on open circuit for half a minute. If the capacitor is not leaky, it will not lose more than 5 or 10 percent of its initial terminal voltage.

Note that the capacitor will normally "soak up" some charge. In other words, if the capacitor terminals are short-circuited, the capacitor will appear to be completely discharged. However, if it then stands for a minute on open circuit, the DVM will show that the capacitor has slowly released some "soaked up" voltage. Thus, the DVM might indicate a terminal voltage of 0.75 volt, or even more.

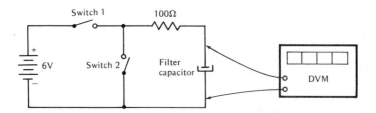

Note: *This is a useful quick check, because many capacitor testers cannot measure the capacitance of a low-voltage high-capacitance electrolytic capacitor.*

The quick check also provides a practical leakage test, as explained in the text.

Figure 14-8   Quick check for large capacitance values.

CHAPTER 15

TROUBLESHOOTING WITH LOGIC PROBES

CHECKING NARROW PULSES • BARRIER POTENTIAL • COINCIDENCE PROBE • ANTI-COINCIDENCE PROBE • AND GATE ACTION • TROUBLE-SHOOTING WITH COMMERCIAL LOGIC PROBES • A COMMON TROUBLE-SHOOTING PROBLEM • FAILURE MODES OF DIGITAL ICs • EFFECT UPON CIRCUIT OPERATION • TOTEM-POLE OPERATION • STEERING CIRCUITRY FAULT

CHECKING NARROW PULSES

Troubleshooters are sometimes concerned with very narrow digital pulses. The basic LED logic-probe configuration shown in Figure 15-1 provides satisfactory indication of logic-high dc levels and of square-wave pulse trains. Adequate light output from the LED is also obtained on pulse trains with comparatively narrow pulse width. As the pulse width is reduced, the light output from the LED decreases. When the probe is applied in a circuit that has very narrow pulses, there is no visible light output from the LED.

However the logic probe can be supplemented with a peak-hold unit, as depicted in Figure 15-2, and very narrow pulses will then produce visible light output from the LED. (The uncharged capacitor has a very low impedance with respect to the circuit under test; it initially draws a heavy pulse current from the circuit, and stores the charge. Then, as the capacitor voltage increases, the capacitor has an increasingly high impedance with respect to the circuit under test.)

By way of comparison, the logic probe has an input resistance of approximately 45 kilohms, and it has no energy-storage capability. Troubleshooting in digital circuits with very narrow pulses is greatly facilitated by a peak-hold unit.

Experiment: The usefulness of this arrangement is easily verified by means of a simple experiment with a square-wave generator. For example, operate the generator at a 6-kHz "clock frequency," and adjust the output to +5V p-p. When the square-wave output is directly applied to the logic probe, the LED glows at comparatively

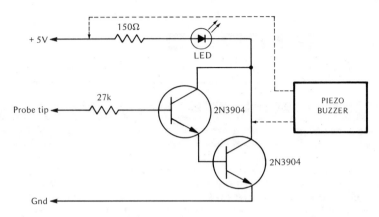

Note: The input resistance of the LED logic probe from the probe tip to ground is approximately 45 kilohms. In other words, the Darlington pair has an input resistance of about 18 kilohms. This is a nonlinear resistance, and its measured value will depend somewhat upon the value of test voltage applied to the probe tip.

Note: For supplementary audio tone indication of a logic-high level, connect a piezo buzzer into the probe circuit, as shown by the dotted lines. The Radio Shack 273-060 solid-state buzzer is suitable.

Figure 15-1   Basic LED logic-probe configuration.

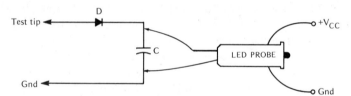

Note: Diode D must have high surge-current capability, and very high reverse resistance. The Radio Shack silicon epoxy rectifier, 276-1114, is suitable. Capacitor C has a value of 4 μF, and must have very high insulation resistance. The GE Pyranol-type capacitor is suitable.

Figure 15-2   Peak-hold unit for logic probe.

high brightness. On the other hand, if the square-wave output is changed into very narrow pulses, as depicted in Figure 15-3, the LED remains dark. Next, if these very narrow pulses are applied to the peak-hold and logic-probe arrangement shown in Figure 15-2, the LED glows visibly.

Follow-up experiment: Now change the 0.001-μF capacitor depicted in Figure 15-3 to a 0.0001-μF capacitor. The resultant output pulses are ten times narrower than before. When these extremely narrow pulses are applied to the peak-hold and logic-probe arrangement shown in

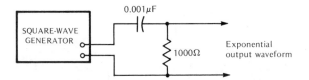

Note: *The square-wave generator used in this experiment should have a fast rise time. The differentiating circuit has a time-constant of 1 microsecond. In turn, the exponential output waveform (a train of exponential pulses) provides a pulse width of 0.7 microsecond.*

Next, when the 0.001 μF capacitor is replaced with a 0.0001 μF capacitor, the differentiating circuit has a time constant of 0.1 microsecond. The output pulse width then is 0.07 microsecond (70 nanoseconds).

Figure 15-3   Square wave is differentiated into very narrow pulses.

Figure 15-2, the LED remains dark. (The LED probe is loading the peak-hold unit sufficiently to prevent a significant charge being built up by the extremely narrow pulses.) Therefore, the troubleshooter proceeds as follows: While the extremely narrow pulses are being applied to the peak-hold unit, the troubleshooter temporarily disconnects the LED probe, thereby eliminating its loading effect. In turn, a substantial charge builds up on the capacitor. Then, when he touches the LED probe tip to the charged capacitor, the LED flashes brightly for a moment, and then goes dark.

The foregoing experiments clearly illustrate how to troubleshoot digital circuits with very narrow pulses (or extremely narrow pulses), using simple and inexpensive test equipment.

BARRIER POTENTIAL

The silicon epoxy rectifier used in the peak-hold unit, like all diodes, has a certain barrier potential. In other words, the diode cannot pass any forward current until the input voltage rises above the barrier potential. Thus, the peak-hold capacitor can never charge up to the full value of the applied input voltage.

Experiment: The barrier voltage of any diode used in a peak-hold unit can be easily measured, as shown in Figure 15-4. A 6-volt battery is connected to the input terminals of the peak-hold unit, then the battery voltage and the capacitor voltage are measured with a DVM. In this example, the battery measures 6.11 volts, and the capacitor measures 5.84 volts. The difference between these two values, 0.27 volt, is the barrier potential of the silicon epoxy rectifier.

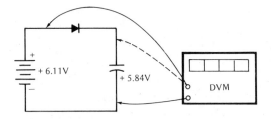

Note: *This experiment illustrates the fact that a peak-hold unit cannot charge up to the full peak value of the applied input voltage. However, in digital troubleshooting procedures, this small loss in voltage is "traded off" for essential charge storage in the capacitor. In turn, although there is a small loss of peak input voltage, the end result is that pulse tests can be made that are far beyond the capability of an unassisted LED logic probe.*

Figure 15-4   Measurement of barrier potential in peak-hold unit.

COINCIDENCE PROBE

Troubleshooters occasionally need to determine whether pulses at one terminal are coincident with pulses at another terminal, or whether the pulses may be anticoincident. For example, a clocked terminal may be in phase with another clocked terminal, or, it may be 180 degrees out of phase with the other clocked terminal.

Pulse coincidence tests are easily made with a coincidence probe such as that shown in Figure 15-5. This arrangement consists of an AND gate, followed by the basic LED logic probe previously described. The AND gate can be placed in the same housing with the probe, and can be "piggy-back" powered from the same +5V probe lead.

Observe that the LED remains dark unless both Input 1 and Input 2 of the AND gate are driven logic-high simultaneously. This is the basis of the pulse-coincidence test. Note that this coincidence probe functions as a basic logic probe if Input 1 and Input 2 are tied together. In this situation, the AND gate serves as a buffer.

Note also that the coincidence probe has a lower input resistance than does a basic logic probe. In other words, it has a greater loading effect on the circuit under test. However, this loading is usually tolerable—it corresponds to 1 unit load (additional fan-out of 1), and most digital circuits are not significantly disturbed by the additional unit load.

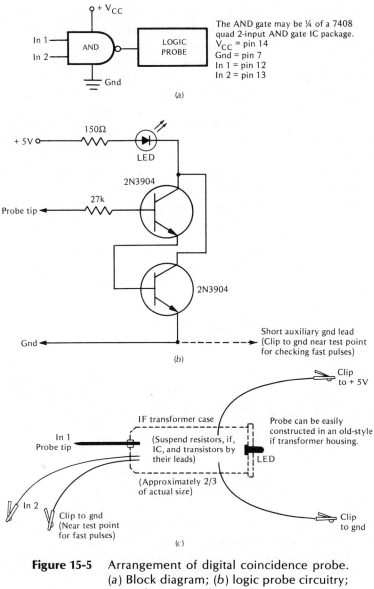

The AND gate may be ¼ of a 7408
quad 2-input AND gate IC package.
V_{CC} = pin 14
Gnd = pin 7
In 1 = pin 12
In 2 = pin 13

(a)

(b)

(c)

Figure 15-5 Arrangement of digital coincidence probe.
(a) Block diagram; (b) logic probe circuitry;
(c) probe construction.

ANTI-COINCIDENCE PROBE

If the troubleshooter wishes to have an anticoincidence probe
available, this device can also be easily constructed in the same

general manner as the coincidence probe. With reference to Figure 15-5, an XOR gate is employed instead of an AND gate. The XOR gate may be ¼ of a 7486 quad 2-input XOR gate IC package.

Observe that the LED in an anticoincidence probe remains dark unless Input 1 and Input 2 are at opposite logic levels. In other words, the LED remains dark if both inputs are logic-high, or if both inputs are logic-low. On the other hand, when Input 1 is driven logic-high at the same time that Input 2 is driven logic-low, the LED glows. Similarly, when Input 1 is driven logic-low at the same time that Input 2 is driven logic-high, the LED glows.

Important: Remember that both inputs of a coincidence probe or an anticoincidence probe must be connected to something if a meaningful indication is to be obtained. In other words, "floating inputs" are effectively logic-high insofar as LED indication is concerned.

Note also that the anticoincidence probe functions as a basic logic probe if Input 2 is connected to gnd. In this situation, the XOR gate serves as a buffer.

AND GATE ACTION

Circuit action for a typical TTL AND gate when both inputs are driven logic-high is shown in Figure 15-6(a). The Q1 emitter junctions are cut off, with the result that Q6 conducts and Q5 is cut off. When both inputs are driven logic-low as shown in (b), the Q1 emitter junctions are turned on, with the result that Q6 cuts off and Q5 conducts.

TROUBLESHOOTING WITH COMMERCIAL LOGIC PROBES

Commercial logic probes, such as that illustrated in Figure 15-7, provide comparatively sophisticated digital test facilities. Thus, the Model 545A localizes nodes stuck high or stuck low, intermittent pulse activity, and normal pulse activity. In other words, the probe shows whether the node being probed is logic-high, logic-low, bad level, open-circuited, or pulsing. The high input impedance of the probe ensures against excessive circuit loading, not just in the logic-high state, but in the logic-low state as well.

The illustrated probe also provides switch-selectable, multifamily operation, and built-in pulse memory. It employs one-lamp display

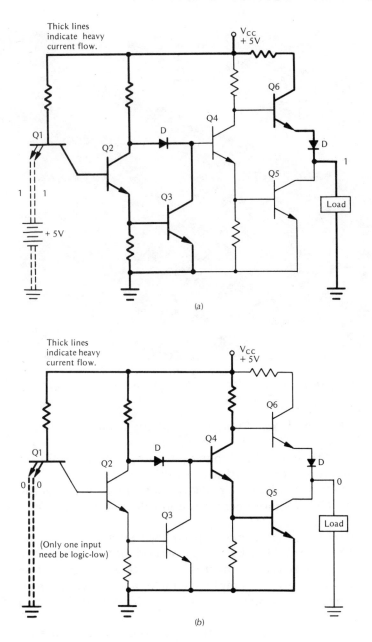

Thick lines indicate heavy current flow.

V_{CC} + 5V

(a)

Thick lines indicate heavy current flow.

V_{CC} + 5V

(Only one input need be logic-low)

(b)

Figure 15-6 AND gate circuit action. (a) Inputs are logic-high; output is logic-high; (b) inputs are logic-low; output is logic-low.

Figure 15-7   View of the Hewlett-Packard logic probe.
(Courtesy, Hewlett-Packard.)

and operates from 4 to 18 volts in CMOS applications, or from 4.5 to 15 volt dc supplies in the TTL mode, while maintaining standard TTL thresholds. The probe's independent, built-in pulse memory assists in capturing intermittent hard-to-see pulses. To catch a glitch, connect the probe to a circuit point, reset the memory, and wait for the glitch to occur. (The memory captures and retains a random pulse until reset.)

Operation at frequencies up to 80 MHz in TTL, and up to 40 MHz in CMOS, is provided by the illustrated probe. Note that when troubleshooting circuits consist of analog components, the task is one of verifying relatively simple characteristics, such as resistance, capacitance, or turn-on voltages of components with two, or at most three nodes.

Although the function of the total circuit may be quite complex, each component in the circuit performs a relatively simple task, and proper operation is easily verified. For example, each diode, resistor, capacitor, and transistor can be tested using a signal generator, voltmeter, ohmmeter, diode checker, or oscilloscope—the traditional troubleshooting tools. *On the other hand, when a circuit is fabricated in IC form, these components are no longer accessible.* It now becomes necessary to test the operation of the complete circuit function.

Therefore, an important difference between discrete circuitry and today's circuits built from digital ICs is in the *complexity of the functions performed by these new "components."* Unlike a resistor, capacitor, diode, or transistor, which must be interconnected to form a circuit function, today's digital IC performs complete, complex functions. Instead of observing simple characteristics, *it is now necessary to observe complex digital signals and to decide if these signals are correct according to the function that the IC is intended to perform.*

Verifying proper component operation now requires stimulating (driving) and observing many inputs (for example, as many as 10 inputs), while simultaneously observing several outputs (often 2 or 3 and at times as many as 8). Another fundamental difference between circuitry built from discrete components and digital ICs is the number of inputs and outputs associated with each component, and the need to stimulate then observe these inputs and outputs simultaneously.

In addition to the problems of simultaneity of signals and complexity of functions at the component level, the digital IC has introduced a new degree of complexity at the circuit level. Circuits which perplex all but their designer are commonplace. If the troubleshooter takes sufficient time, he can study these circuits and understand their operation. However, this is not an affordable luxury for the busy digital technician. Accordingly, without understanding a circuit's intricate operation, *it becomes necessary to have a technique of quickly testing each component, rather than attempting to isolate a failure to a particular circuit segment by testing for expected signals.*

Therefore, in order to solve these problems and to make the troubleshooting of digital circuits more efficient, it is necessary to take advantage of the digital nature of the signals that are involved.

Refer to Figure 15-8. This is an example of a typical TTL signal. An oscilloscope displays *absolute voltage* with respect to time, but in the digital domain absolute values are unimportant. A digital signal exists in one of two or three states—high (true), low (false), and undefined or in-between level—each of which is determined by a threshold voltage. It is the relative value of the signal voltage with respect to these thresholds that determines the state of the digital signal, and *this digital state determines the operation of the IC*—not absolute levels.

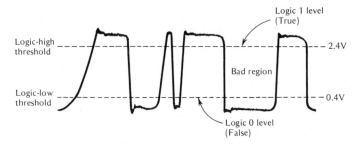

Note: In the digital domain, the relative value of a signal voltage with respect to the threshold voltages *determines the operation of the circuit. A signal above the high threshold is in the high state, and whether it is 2.8 volts, or 3.0 volts, is unimportant to the operation of the circuit.*

Figure 15-8   Example of a TTL signal waveform.
(Courtesy, Hewlett-Packard.)

Note that in Figure 15-8, if the signal is greater than 2.4 volts, it is in a high state—it is unimportant whether the level is 2.8 or 3.0 volts. Similarly, for a low state, the voltage must be below 0.4 volt. *It is not important what the absolute level may be, as long as it is below this threshold.* Thus, when using an oscilloscope, the troubleshooter must determine over and over again whether the signal meets the threshold requirement for the discrete digital state—a needless waste of time.

Within a digital logic family, the timing characteristics of each component are well defined. Each gate in the TTL family displays a characteristic *propagation delay time, rise time, and fall time.* The effects of these timing parameters on circuit operation are taken into account by the designer. Once a design has been developed beyond the breadboard or prototype stage and is into production, problems due to design have ordinarily been corrected.

An important characteristic of digital ICs is that when they fail, they fail catastrophically. This means that the timing parameters rarely degrade or become marginal. Thus, the troubleshooter who observes waveforms on an oscilloscope and makes repeated decisions concerning the validity of timing parameters is wasting time and is contributing very little to the troubleshooting process.

Once problems due to design are corrected, the fact that pulse activity exists is usually sufficient indication of proper IC operation without further observation of pulse width, repetition rate, rise time, or fall time.

A COMMON TROUBLESHOOTING PROBLEM

Troubleshooters frequently encounter the problem of "stuck low" in TTL circuitry. For example, with reference to Figure 15-9, the output stage of the TTL device is a transistor totem pole. In either the high or the low state, it presents a low impedance. In the low state, it is a saturated transistor to ground. Thus, it "looks like" a 5 or 10 ohm resistance to ground.

This low impedance presents a problem to in-circuit stimulation. In other words, a signal source that is used to inject a pulse at a node which is driven by a TTL output must have sufficient power to override the low impedance output state. *Ordinary sources used for trouble-shooting do not have this capability.* It has been necessary in the past for the troubleshooter to either cut printed-circuit traces, or to "pull" IC leads in order to stimulate the circuit under test. Both of these procedures are time-consuming and lead to unreliable repairs.

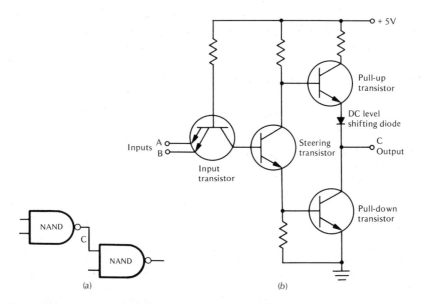

Note: *When stimulating a node in-circuit, such as C (above), it is necessary to override the low-impedance totem-pole output stage, which is driving the node. When the output is in a low state, it is a saturated transistor to ground. (Ordinary signal sources are not powerful enough to override this low state.)*

Figure 15-9   Totem-pole output stage. (a) Logic diagram; (b) configuration.

Use of the traditional oscilloscope and traditional signal sources is an inefficient approach. These troubleshooting tools are of the general-purpose type that can be applied in any situation, provided that the troubleshooter has sufficient time. But, with the quantity and complexity of today's electronic circuits, it is essential to find the most efficient solution to the problem at hand.

These considerations lead to the conclusion that the oscilloscope, diode checker, and voltmeter should be reserved for troubleshooting analog circuits, where they really shine, and to use instruments that take advantage of the digital nature of signals for troubleshooting digital circuitry.

FAILURE MODES OF DIGITAL ICs

To troubleshoot digital circuits efficiently, it is essential to understand the types of failures that occur. These failures can be catagorized into two main groups:

1. Faults internal to the IC.
2. Faults in the circuit external to the IC.

These are four types of defects that can occur internally to an IC, as follows:

1. An open bond (connection) on either an input or an output.
2. A short-circuit between two pins (neither of which are V_{CC} or ground).
3. A short-circuit between an input or an output and V_{CC} or ground.
4. A failure in the internal circuitry (often called the steering circuitry) of the IC. (See Chart 15-1.)

In addition to these four failures internal to an IC, there are four faults that can occur in the circuit external to the IC. These are:

1. A short-circuit between a node and V_{CC} or ground.
2. A short-circuit between two nodes (neither of which are V_{CC} or ground).
3. An open signal path.
4. A failure of an analog component.

EFFECT UPON CIRCUIT OPERATION

Consider the effect that each of the foregoing failures has upon circuit operation. The first failure internal to an IC mentioned above

Chart 15-1

___

TOTEM-POLE OPERATION

A NAND gate schematic is shown in Figure 15-10. Thick lines (Figure 15-11) denote heavy current flow. When the inputs of transistor Q1 are logic-low (grounded), Q1 becomes forward-biased and conducts heavily (is saturated). The current flow is then diverted away from Q2 and flows to ground.

Accordingly, Q2 cuts off and its collector resistance becomes very high. The base of Q3 becomes forward-biased, and Q3 conducts heavily (is saturated). In turn, current flows from the emitter of Q3, through diode D, and into the output line; the output lines go logic-high. Note that Q4 is cut off; therefore, Q4 cannot shunt any current to ground.

Next, when the inputs of transistor Q1 are logic-high (at V_{CC} potential) (as shown in Figure 15-12), the base-emitter junction is reverse-biased and cannot conduct any current. On the other hand, the base-collector junction of Q1 is then forward-biased and a heavy current flows into the base of Q2. In turn, Q2 becomes saturated and has a very low base-emitter resistance.

This condition permits a heavy current flow from the emitter of Q2 into the base of Q4, and from the emitter of Q4 to ground. The heavy base-emitter current in Q2 diverts current flow from the base of Q3, with the result that Q3 is cut off and has a very high collector-emitter resistance.

Because Q4 is saturated, its collector-emitter resistance is very low. Therefore, the output line goes logic-low. Note that diode D is included in the emitter branch of Q3 to provide a small additional voltage drop which ensures that Q3 will be completely cut off.

Faults in the steering circuitry (phase splitter) include collector-base leakage, open collector, open base, base-emitter leakage, and open emitter. A steering-circuit defect results in a "stuck at" trouble symptom.

___

was an open bond on either an input or on an output. This failure will have a different effect, depending upon whether it is an open output bond or an open input bond.

In the case of an open output bond (gate output lead broken inside of the IC package), the inputs of following devices that are driven by that output are left to "float." In TTL and DTL circuits, a floating input rises to approximately 1.4 to 1.5 volts, and usually has the

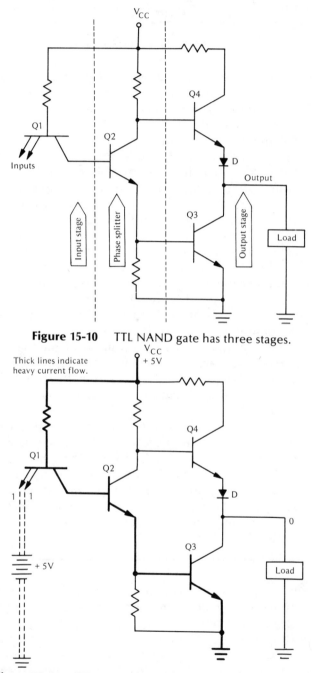

Figure 15-10   TTL NAND gate has three stages.

Figure 15-11   TTL NAND gate; logic-low input produces logic-high output.

same effect on circuit operation as a high-logic level. Thus, an open output bond will cause all inputs driven by that output to float to a bad level, since 1.5 volts is less than the high threshold level of 2.0 volts, and greater than the low threshold level of 0.4 volt. (In TTL and DTL, a floating input is interpreted as a high level. Thus, the effect will be that these inputs will respond to this bad level as though it were a static-high signal.)

In the case of an open input bond (gate input lead broken inside of the IC package), the troubleshooter will find that the open circuit blocks the signal driving the input from entry into the IC chip. In other words, the input on the chip is allowed to float, and the device will respond as if this input were a static-high signal.

Important: Since the open-circuit under discussion occurs on the input inside of the IC, the digital signal driving this input will be unaffected by the open circuit, and it will be detectable when "looking at" the input pin to the internal defect. The effect, then, will be to block this signal inside the IC, and the resulting IC operation will be as though the input were a static high.

A short-circuit between an input and V_{CC} or ground has the effect of holding all signal lines connected to that input (or output) either

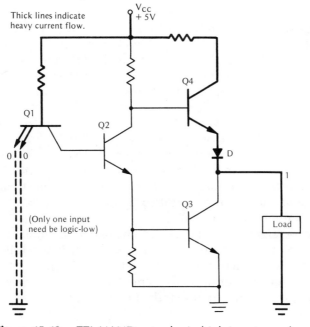

Figure 15-12   TTL NAND gate; logic-high inputs produce
logic-low output.

high, as in the case of a short to V_{CC}, or low, as in the case of a short to ground. (The short-circuit might be inside the IC, or it might occur in the external circuit.) In many cases, this will cause expected signal activity beyond the short-circuit point to disappear, and hence this type of failure is catastrophic in terms of circuit operation.

A short-circuit between two pins of an IC is not as straightforward to analyze as a short to V_{CC} or to ground. When two pins are shorted, the outputs driving those pins oppose each other; when one output attempts to pull the shorted pins high, the other output attempts to pull the shorted pins low.

In this situation, the output attempting to go high will supply current through the upper saturated transistor of its totem-pole output stage, while the output attempting to go low will sink this current through the lower saturated transistor of its totem-pole output stage. The bottom line is that *the short will be pulled to a low state by the saturated transistor to ground.* Whenever both outputs attempt to go high simultaneously, or to go low simultaneously, the shorted pins will respond normally. *But whenever one output attempts to go low, the short will be constrained to the low state.*

STEERING CIRCUITRY FAULT

The fourth failure internal to an IC is a defect of the internal (steering) circuitry of the IC. This defect permanently turns on either the upper transistor of the output totem pole (Figure 15-9) and locks the output in the high state, or turns on the lower transistor of the totem pole and locks the output in the low state. This failure blocks signal flow and has a catastrophic effect on circuit operation.

A short-circuit between a node and V_{CC} or ground external to the IC is indistinguishable from a short internal to the IC. Both defects will cause the signal lines connected to the node to be either always high (for shorts to V_{CC}) or to be always low (for shorts to ground). When this type of failure is encountered, the troubleshooter must make a very close examination of the circuit to determine whether the fault is internal to the IC, or external.

An open signal path in the circuit has an effect similar to an open output bond driving the node. In other words, all inputs following the open circuit will float to a bad level, and they will appear as static high levels in circuit operation. On the other hand, all inputs preceding the open circuit will be unaffected and will respond as in normal operation.

CHAPTER 16

TROUBLESHOOTING WITH LOGIC PULSERS AND PROBES

PULSE TRAINS VS. STATIC-HIGH LEVEL • *HOW TO MONITOR INTERMITTENTS WITH LOGIC-HIGH AND LOGIC-LOW PROBES* • *PULSE MEMORY IN COMMERCIAL PROBES* • *BASIC STEPS IN DIGITAL TROUBLESHOOTING* • *LOGIC PULSER OPERATION* • *CAUSES OF SHORT-CIRCUITS* • *TTL VS. CMOS TESTING PROCEDURES* • *FAILURE MODES IN DIGITAL CIRCUITRY* • *EXPERIMENT*

PULSE TRAINS VS. STATIC-HIGH LEVEL

Pulse trains produce the same general type of indication as a static logic-high level, when checked with a basic logic probe. (If a slow pulse generator is used, the LED will flicker; however, at high pulse repetition rates, persistence of vision causes the appearance of a steady glow from the LED.)

When the troubleshooter needs to distinguish between a pulse train and a static-high level, *a blocking capacitor may be utilized in series with the probe input,* as shown in Figure 16-1. The test results are as follows:

1. When a steady dc voltage is applied to the blocking capacitor, the LED will flash momentarily, and then remain dark. (A momentary pulse current is drawn as the blocking capacitor charges.)
2. When a digital pulse train is applied to the blocking capacitor, the LED will glow steadily, although at reduced brightness in comparison with direct application of the pulse train to the probe tip. (The brightness is reduced because the dc component of the waveform is rejected. Also, at low repetition rates, the pulse is partially differentiated.)

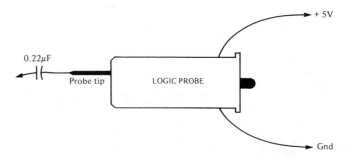

Note: When a dc source is applied to the 0.22 μF capacitor, the LED flashes briefly and then remains dark. On the other hand, when a train of digital pulses is applied to the 0.22 μF capacitor, the LED glows steadily. Thus, the blocking capacitor serves to show whether the point under test is a logic-high dc level, or whether it is being pulsed high by a train of pulses.

If the probe tip is directly applied to a dc source, the LED glows steadily. If the probe tip is applied directly to train-of-pulses source, the LED will glow steadily, although at reduced brightness. The blocking capacitor thus serves to "prove" that the source is either steady dc, or a pulse train.

Figure 16-1 Blocking capacitor assists troubleshooter in distinguishing between a pulse train and a steady dc level.

Increasing the Light Output

A helpful trick of the trade in the foregoing test procedure consists of using a separate 9-volt battery to power the probe instead of a "piggy back" connection to V_{CC} in the digital system. The 9-volt supply for the probe compensates for the lower conduction level of the Darlington pair in the probe. This arrangement is shown in Figure 16-2.

HOW TO MONITOR INTERMITTENTS WITH LOGIC-HIGH AND LOGIC-LOW PROBES

Intermittents are the curse of the troubleshooter, owing to the fact that an intermittent condition occurs at unpredictable intervals. Troubleshooting of intermittents is facilitated by monitoring key test points with logic-high and logic-low probes with tone indicators. Provision of high-low tone indication permits the troubleshooter to occupy himself elsewhere while awaiting the possible onset of an intermittent condition.

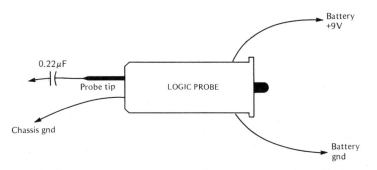

Note: To obtain a brighter glow from the LED when a blocking capacitor is used with the probe, connect the probe to a 9-volt battery, as shown. (The reason that the LED glows with less brightness when a blocking capacitor is used in series with the probe is rejection of the dc component in the digital pulses by the blocking capacitor. This loss of the dc component results in effective reduction in energy content of the pulse train.)

Figure 16-2   Higher probe-battery voltage is used for increased brightness.

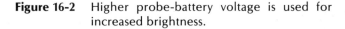

A logic-high probe with tone indicator was depicted in Figure 15-1. With reference to Figure 16-3, a logic-low probe with tone indicator employs a shunt connection of the indicator, instead of a series connection. If the piezo buzzer types noted in the diagram are utilized, a logic-high state will be indicated by a high audio tone, and a logic-low state will be indicated by a low audio tone. To monitor a test point for intermittent operation, connect both the logic-high probe and the logic-low probe to the test point. Indication is as follows:

1. If the test point is logic-high, a steady high audio tone is provided.
2. If the test point is logic-low, the high audio tone stops and a low audio tone takes its place.

PULSE MEMORY IN COMMERCIAL PROBE

A commercial logic probe, such as the Hewlett-Packard 545A described in Chapter 4, contains a lamp indicator. The lamp glows brightly in response to a logic-high level, goes dark in response to a logic-low level, and glows dimly in response to a "bad" level.

Since the troubleshooter must observe dynamic signal activity, as well as the static levels described above, a commercial logic probe also provides pulse-stretching circuitry which can detect pulses as narrow

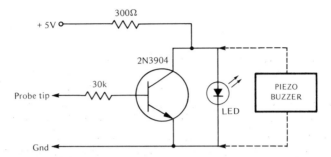

Note: *The LED may need to be selected for complete darkness when the probe tip is logic-high. If the probe tip is logic-low, the LED glows; it also glows if the probe tip is "floating." The piezo buzzer provides audible logic-low indication; it may be a Mallory 5C628.*

Figure 16-3 Logic-low probe, used for intermittent monitoring.

as 10 ns, and stretch them to produce a 0.1 second blink of the lamp indicator. Accordingly, if a low signal pulses high, the indicator lamp will blink "on"; if a high signal pulses low, it will blink "off."

A *pulse memory* option is provided with this type of probe to monitor a signal line for single-shot or low-frequency pulses over extended periods of time. Upon the occurrence of a pulse, an LED in the pulse memory will glow, and it will continue to glow until it is reset by the troubleshooter.

BASIC STEPS IN DIGITAL TROUBLESHOOTING

Professional troubleshooters know that the first step in any troubleshooting process is to narrow the malfunctioning area as much as possible by examining the observable characteristics of the failure. This is often called "front-panel milking." From the front-panel operation (or rather, mis-operation) the failure should be localized to as few circuits as possible. From this point, it is necessary to further narrow the failure to one suspected circuit by looking for improper key signals between circuits. A logic probe can be very effective in this procedure.

In many cases, a signal will completely disappear. By probing the interconnecting signal paths, a missing signal can be readily detected. Another important failure is the occurrence of a signal on a line that should not have had a signal. A pulse memory function for a logic probe enables such signal lines to be monitored for single-shot pulses or pulse activity over extended periods of time. (The occurrence of a signal will be stored and indicated on the pulse memory's LED.)

Dependence upon a well-written service manual is the key to this phase of troubleshooting. Isolating a failure to a single circuit requires knowledge of the instrument or system and its operating characteristics. A well-written manual will indicate the key signals to be observed. *The logic probe provides a rapid means of observing the presence of these signals.*

LOGIC PULSER OPERATION

We have previously considered elementary pulsing procedures, such as touching a test lead to V_{CC}, or charging a fixed capacitor to V_{CC} and then discharging the capacitor into the circuit point which is to be pulsed. *The mainstay of all digital troubleshooting techniques is stimulus-response testing.* In other words, it is necessary to apply a test signal, and to observe the resulting response. In turn, the trouble-shooter can conclude whether the device is operating properly.

A logic pulser, such as that illustrated in Figure 16-4, provides a professional source of digital test pulses. The logic pulser injects into the circuit under test a single 300 ns wide pulse of proper amplitude and polarity each time its button is pressed. (See Figure 16-5.) If the node was low, it will be pulsed high—if it was high, it will be pulsed low, without readjusting the pulser.

The illustrated pulser is capable of sourcing or sinking 0.75A for the 300 ns pulse width to ensure that the node changes state. (The narrow width of 300 ns avoids damage to the ICs being pulsed.) The troubleshooter may now move rapidly from point to point in the circuit under test by applying pulses and observing the resulting responses.

The first failure to test for is an open bond in the IC driving the failed node. A logic probe provides a quick and accurate test in this situation. If the output bond is open, then the node will float to a bad level. By probing the node, the logic probe will quickly indicate a bad level. In the event that a bad level is indicated, the IC driving the node should be replaced.

If the node is not at a bad level, then a test for a short-circuit to V_{CC} or ground should be made. This can be easily done with the logic pulser and probe. While the logic pulser is sufficiently powerful to override even a low-impedance TTL output, it is not powerful enough to effect a change in state on a V_{CC} or ground bus. Thus, if the logic

Figure 16-4   A professional logic pulser.
(Courtesy, Hewlett-Packard.)

pulser is used to inject a pulse while the logic probe is applied simultaneously on the same node to observe the pulse, a short to V_{CC} or ground can be detected. *The occurrence of a pulse indicates that the node is not shorted, and the absence of a pulse indicates that the node is shorted to V_{CC} (if it is logic-high) or to ground (if it is logic-low).*

CAUSES OF SHORT-CIRCUITS

If a node is short-circuited to V_{CC} or to ground, there are two possible causes. The first possibility is a short in the circuit external to the ICs, and the other possibility is a short internal to one of the ICs connected to the node. An external short should be localized by an examination of the circuit. *If no external short is found, then the cause is equally probable to be any one of the ICs connected to the node.*

An educated guess that can be made (based upon experience) is to first replace the IC driving the node, and if that does not solve the problem, to try each of the other ICs individually until the short is

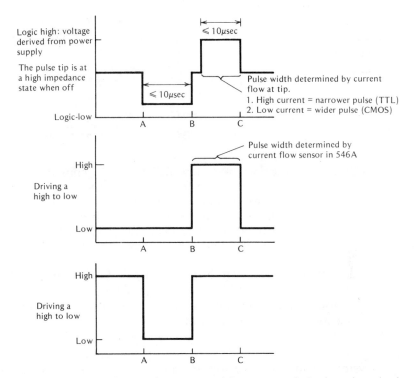

Logic high: voltage derived from power supply

The pulse tip is at a high impedance state when off

≤ 10μsec

≤ 10μsec

Logic-low

A   B   C

Pulse width determined by current flow at tip.
1. High current = narrower pulse (TTL)
2. Low current = wider pulse (CMOS)

Pulse width determined by current flow sensor in 546A

High

Driving a high to low

Low

A   B   C

High

Driving a high to low

Low

A   B   C

Note: *The H-P logic pulser can be programmed to output a desired number of pulses, instead of a single-shot pulse. For example, if the troubleshooter wishes to obtain 432 pulses, he presses and releases the code button and the latch code button for programming four 100-Hz bursts, three 10-Hz bursts, and two single pulses.*

Figure 16-5   Pulse waveforms outputted by the H-P logic pulser.
(Courtesy, Hewlett-Packard.)

eliminated. (It can occasionally happen that analog components such as resistors or capacitors connected to the node have shorted.) An example of a solder bridge short-circuit is depicted in Figure 16-6.

If the node is not shorted to V_{CC} or to ground and if it is not an open output bond, then the troubleshooter looks for a short between two nodes. This is accomplished in one of two ways:

1. The logic pulser can be used to pulse the failed node under test, and the logic probe can be used to observe each of the remaining failed nodes. If a short exists between the node being checked and one of the other failed nodes, then the pulser will cause the node being probed to change state (i.e.,

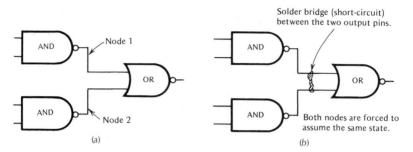

Figure 16-6   Example of a "stuck-low" trouble symptom.
(a) Node 1 can be driven high, and node 2 can
be driven low by the AND gates; the OR-gate
output goes high; (b) if there is a short-circuit
between the two AND-gate output pins, the
OR-gate output is "stuck-low."

the probe will detect a pulse). To verify that a short exists, the
probe and pulser should be reversed and the test made again.
If a pulse is again detected, then the troubleshooter concludes
that a short is certainly present.

2. As a further test, or as another way of testing for a short
between two nodes, the circuit can be removed from the
system or instrument, and an ohmmeter may be applied to
measure the resistance between the two failed nodes.
Thereby, a short-circuit between them can be easily
determined.

If the failure is a short-circuit, there are two possible causes. The
most likely cause is a problem in the circuit external to the ICs. This
condition can be localized by examining the circuit and repairing any
solder bridges or loose-wire short-circuits that are spotted. Only if the
two nodes that are shorted are common to one IC can the failure be
internal to that IC. If, after examining the circuit, no short can be
found external to the IC, then the IC should be replaced.

If the failure is not a short between two nodes, there are only two
possibilities left. These are: an open input bond, or a failure in the IC
internal circuitry. In either case, the IC should be replaced. In other
words, the cause of a failed node is determined by systematically
eliminating the IC failures.

One type of failure that has not been discussed is an open signal
path in the circuit external to the IC (Figure 16-7). If, after correcting
all of the defects that can be located by the techniques explained
above, the circuit is still found to be malfunctioning, then the
troubleshooter will suspect that an open signal path is present.

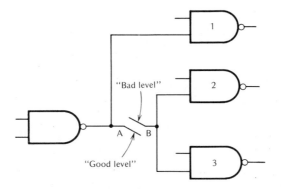

Note: *The open circuit in this example causes point B to float to a bad level, while point A is driven by proper TTL or DTL signal levels. The troubleshooter starts at the input of gate 3 and proceeds back toward gate 1 to determine the exact location of the open, using the logic probe.*

Figure 16-7 Example of an open signal path in the circuit external to the ICs.
(Courtesy, Hewlett-Packard.)

A logic probe provides a rapid means, not only of showing the presence of an open circuit, but also of localizing the fault. Inasmuch as an open signal path allows the input to the "right" of the open to float to a bad level, the logic probe can be used to test the input of each IC for a bad level. Then, after an input floating at a bad level is detected, the logic probe can be used to follow the circuit back from the input to search for the open.

This procedure is effective because the circuit to the "left" of the open will display a good logic level (either high, low, or pulsating) while the circuit to the right will display a bad level. Thus, probing back along the signal path will indicate a bad level until the open is passed. Then, the probe can precisely locate the open.

To repeat a basic principle, a logic pulser and logic probe provide two chief types of troubleshooting data: Stimulus/response testing at the gate level, and truth-table verification. (See Figure 16-8.) As explained above, the pulser and probe also are useful in localizing external circuit defects.

TTL VS. CMOS TESTING PROCEDURES

Although the troubleshooter is usually concerned with TTL configurations, CMOS circuitry will also be encountered. Most logic probes have a switch to accommodate either the TTL family or the

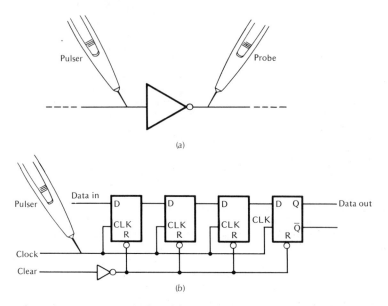

(a)

(b)

Note: These are in-circuit examples of digital test procedures with a logic pulser and a logic probe. In (a) the input of the inverter is pulsed to its opposite prevailing state, and the output is probed to determine whether it changes state in response to the test pulse. In (b) the shift register is clocked by means of the pulser, to determine whether it is following its truth table.

Note: The internal impedance (resistance) of a node is often quite low. Accordingly, the source impedance of a pulser must be less than 2 ohms in order to override the internal impedance of the circuit under test.

Figure 16-8   Examples of basic pulser-probe tests. (a) Stimulus/response testing at the gate level; (b) truth table verification.
(Courtesy, Hewlett-Packard.)

CMOS family. (See Figure 16-9.) Professional-type logic pulsers are also designed to accommodate TTL and CMOS circuitry. For example, the pulser illustrated in Figure 16-4 has the following characteristics:

1. When applied in a TTL circuit with a power-supply voltage from 4.5 to 5.5 volts, the pulser outputs a 3-volt logic-high pulse and an 0.8-volt logic-low pulse. The pulse width is $0.5\,\mu s$, and the pulse current is 650 mA.
2. When applied in a CMOS circuit with a power-supply voltage from 3 to 15 volts, the pulser outputs a logic-high pulse with an amplitude equal to the supply voltage less 1 volt, and a logic-low pulse with an amplitude of 0.5 volt. The pulse width is $5\,\mu s$, and the pulse current is 100 mA.

• Allows TTL/CMOS probing.
  1. Set switch to family under test.
  2. Attach supply leads to power source of family under test.
  3. Select TTL operation using CMOS supply by putting switch in TTL position.

• Catches single pulses.
• Indicates absence of a single pulse.
  1. Plate tip on circuit under test.
  2. Press MEM/CLR; light goes out.
  3. Light comes on when a single pulse occurs.

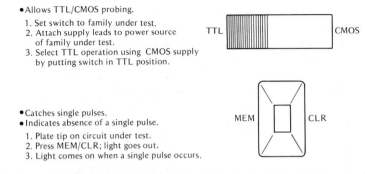

Note: A typical TTL logic-1 level is 2.4 volts, and logic-0 level 0.4 volt. Comparatively, a typical CMOS logic-1 level is 4.2 volts and logic- 0 level 1.8 volts. An occasional CMOS configuration may operate from a 3-volt supply, and another may operate from a 15-volt supply. (A 15-volt supply makes possible a much faster operating speed than a 3-volt supply. TTL is faster than CMOS, but it also has a greater current demand.)

Figure 16-9   Logic probe pulse memory and TTL/CMOS switch.
(Courtesy, Hewlett-Packard.)

When checking an AND gate, for example, the troubleshooter must pulse two or more inputs simultaneously. For this purpose, a logic pulser is used with a multipin stimulus cable assembly. It attaches to the pulser output and fans out into four test leads, each of which is terminated by a micro test clip. Note that when two or more terminals are being simultaneously pulsed, the terminals must all be in the same state (either all logic-high, or all logic-low).

Of course, gates and other logic devices can also be tested out-of-circuit. In this case, note that all unused input must be either tied to a logic-high point, or to a logic-low point, as the test requirement may dictate. Or, an unused input may be tied to some other used input for test purposes. The essential consideration is to avoid any "floating" terminals in a test setup, inasmuch as an unused terminal will "look" logic-high.

Another practical consideration is to avoid slow-rise and/or slow-fall test signals when checking CMOS circuitry. In other words, slow transitions from logic-low to logic-high, or from logic-high to logic-low, result in excessive power dissipation. Professional pulsers avoid this hazard inasmuch as they output fast-rise and fast-fall pulses.

FAILURE MODES IN DIGITAL CIRCUITRY

It is helpful to note the failure modes that occur in digital circuitry, and their statistical order. As seen in Figure 16-10, 75 percent of digital faults can be localized by means of voltage-based troubleshooting instruments. The other 25 percent of digital faults can be localized by means of current-based troubleshooting instruments.

Voltage-oriented faults, in order of frequency, are open inputs, half of which are "stuck open," and open outputs, most of which are "stuck open." Current-oriented faults are chiefly short-circuits, and most of these short-circuits are "stuck nodes." Some of the current-oriented faults involve a substrate input diode short, of which half represent "stuck nodes." Current-oriented short-circuits are subclassified into shorts between traces, shorts to V_{CC}, and shorts to ground.

In most situations (although not in all) there will be only one fault in a malfunctioning circuit; device failures are nearly always catastrophic. The chief exception is intermittent malfunctioning, which will often be localized to a poor connection or poor contact.

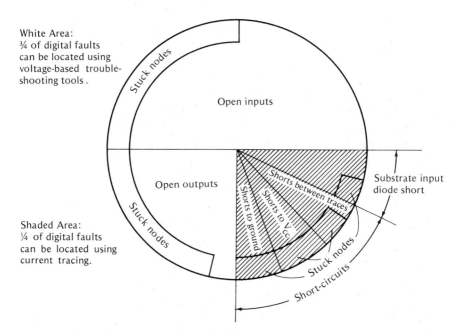

Note: *Nearly all digital failures are catastrophic; devices seldom become marginal (although this can happen on occasion). V_{CC} sometimes becomes marginal, or ripple sometimes becomes abnormal. In turn, devices may "look marginal."*

Figure 16-10   Failure modes and statistical incidences.
(Courtesy, Hewlett-Packard.)

EXPERIMENT

If you are not familiar with shift registers such as that depicted in Figure 16-8(*b*), it will be instructive to experiment with a simple discrete-logic 4-bit shift register, using the flip-flop diagrammed in Figure 16-11. The shift-register arrangement is shown in Figure 16-12. This is an example of discrete logic that is easy to construct. The action of the recirculating shift register can be indicated by means of simple logic probes connected at the flip-flop Q outputs.

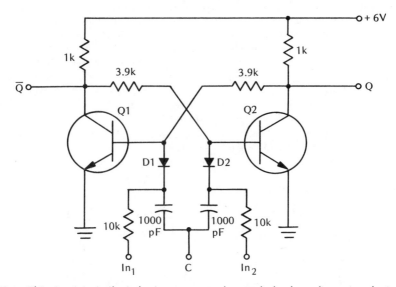

Note: *This circuit is similar in basic respects to the toggle-latch configuration depicted in Figure 13-1. However, this circuit is provided with input terminals In$_1$ and In$_2$. It operates as a clocked toggle flip-flop. (See Figure 16-12.)*

Figure 16-11   Flip-flop configuration used in shift-register experiment.

This binary shift register (left-shift register) will shift (clock) any 4-bit binary number to the left and perform an "end-around" shift from FF4 to FF1. As an illustration, consider the 4-bit binary number 0001. With the occurrence of each clock pulse, the register readout progresses: 0010, 0100, 1000, and then 0001 (starts over). A single-shot pulser may be used to enter or to clear a bit from a flip-flop by pulsing the base of Q1 (or Q2) of a particular flip-flop, while the clock is stopped.

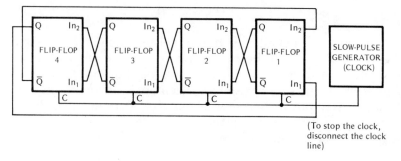

(To stop the clock, disconnect the clock line)

Note: The C, Q, $\overline{Q}$, In$_1$, and In$_2$ terminals refer to the notation in Figure 16-11. The slow pulse generator has the configuration shown in Figure 13-7. If a data bit is entered into flip-flop 1 (as by momentarily pulsing the base of Q1), this data bit will be clocked progressively from FF1 to FF2 to FF3 to FF4 and then back to FF1, from which it will recirculate through the shift register until the register is cleared, and new data entered.

When the shift register is powered-up, its high and low states are unpredictable. However, whatever combination of data bits may appear at power-up will be recirculated until the register is cleared, and new data is entered. (Stop the clock while the register is being cleared or new data is being entered.)

Caution: *When the register is manually cleared, use a test lead with a series 10-kilohm resistor to pulse the base of Q2 (Figure 16-11) from the +6V V$_{CC}$ source. Do not directly connect the base of Q2 to V$_{CC}$. Similarly, when a bit is manually entered into the register, use a test lead with a series 10-kilohm resistor to pulse the base of Q1 (Figure 16-11). Do not directly connect the base of Q1 to V$_{CC}$.*

Figure 16-12   Experimental 4-bit recirculating shift-register arrangement.

INDEX